Automation, Collaboration, & E-Services

Volume 6

Series Editor

Shimon Y. Nof, PRISM Center, Grissom, Purdue University,
West Lafayette Indiana, IN, USA

The Automation, Collaboration, & E-Services series (ACES) publishes new developments and advances in the fields of Automation, collaboration and e-services; rapidly and informally but with a high quality. It captures the scientific and engineering theories and techniques addressing challenges of the megatrends of automation, and collaboration. These trends, defining the scope of the ACES Series, are evident with wireless communication, Internetworking, multi-agent systems, sensor networks, cyber-physical collaborative systems, interactive-collaborative devices, and social robotics – all enabled by collaborative e-Services. Within the scope of the series are monographs, lecture notes, selected contributions from specialized conferences and workshops.

More information about this series at http://www.springer.com/series/8393

Hao Zhong · Shimon Y. Nof

Dynamic Lines
of Collaboration

Disruption Handling & Control

 Springer

Hao Zhong
Facebook Inc.
Menlo Park, CA, USA

Shimon Y. Nof
PRISM Center
Purdue University
West Lafayette, IN, USA

ISSN 2193-472X   ISSN 2193-4738   (electronic)
Automation, Collaboration, & E-Services
ISBN 978-3-030-34465-8   ISBN 978-3-030-34463-4   (eBook)
https://doi.org/10.1007/978-3-030-34463-4

This Springer imprint is published by the registered company Springer Nature Switzerland AG
The registered company address is: Gewerbestrasse 11, 6330 Cham, Switzerland

To our wives, Xiao Ge
and Nava C. Vardinon Nof

Preface I

Most of the time, to most of us, the word *Disruption* implies that something bad has happened: a breakdown, a loss, a mishap, or worse yet, a disaster. Sometimes, we hear: "Please don't disrupt me, let me finish my sentence!"

Between the range of those examples, there are also positive disruptions. Our three favorite ones are:

- Computers! They disrupted, transformed, and continue to transform the world. We remember the early days of computers in the previous century, when some people disliked computers. In response, Isaac Asimov, our dear friend, said: "I do not fear computers. I fear the lack of them!" By now, some people still may claim that they dislike computers. But as soon as they pick up their cellphone… and when you observe how they are glued to its screen…they clearly like and appreciate this helpful, personal, hand-held powerful computer. Yes, when our cellphone is disrupted, we meet the more common, negative disruption.
- Robots! They have disrupted our world of work, of automation, of security, and what not? There is still a fear of robots. To quote from the introduction to our *Handbook of Industrial Robotics*: "Robots should replace people on jobs robots can do more economically. This will initially disadvantage many, but inevitably will benefit all…" By now we know that robots are also essential for disruption handling and control, as described in this book.
- The cute stickers with which some produce growers adorn melons, clementines, and other fruits. These friendly stickers have disrupted the fruit market: They are meant to make us happy and smile; although some competitor growers may frown on the idea.

The problem is that there are many dangerous, unfriendly, and even disastrous disruptions. This book is devoted to their handling and control, with particular focus on the recently developed and powerful DLOC models of dynamic lines of collaboration.

Menlo Park, CA, USA   Hao Zhong
West Lafayette, IN, USA   Shimon Y. Nof
September 2019

Preface II

Early in this century, the Indiana Department of Transportation (INDOT) challenged our PRISM Center at Purdue University to research and develop an essential frontier solution (at that time). INDOT leaders asked: Are we able to R&D transportation safety and security decision-support tools and simulators for training, and for emergency support in real-time, during disruptions to transportation, such as natural calamities and man-made disasters?

We had extensive experience in this area for manufacturing, production, and logistics automation, and for industrial robotics. It was developed in the context of CCT, the Collaborative Control Theory, which was emerging around that time. In other projects, we gained expertise in intelligent transportation. In all these projects, our PRISM researchers combined cybernetics, artificial intelligence, operations research, human factors, and systems engineering, to design and implement cyber-augmented models and workflow task-administration protocols for collaborative e-Work and robotics. Many of those systems included humans and automation.

A relevant example was our original work on the CCT-based design of multi-agent systems and protocols for disruptive errors and conflicts detection, prediction, prevention, and recovery. It was created and implemented with network science foundation for several robotics and service facilities (some simple, some complex); and for concurrent design of sensor networks, supply chains, and manufacturing facilities. So we responded that we were ready and eager to oblige.

Over several years, new specific solutions have evolved and been refined through interactive presentations, validation "mock drills," and workshops for field-testing the simulators, knowledge bases, adaptive and learning communications mechanisms, and responsive interfaces. Users' experiences (then called user interface, and human–decision portals interactions) were carefully designed and tested in periodic workshops with INDOT's researchers, state-wide personnel and first responders, including those from other agencies' security and emergency responders.

Our immense learning experience from this era, working closely with all the many INDOT and related participants (and probably similar to yours, the readers, in similar projects), was both heart-warming and educational:

- Heart-warming, because the participants were evidently compassionate and caring, and one could easily recognize that they knew well, and were serious about their safety and security roles and responsibilities. Moreover, we, as engineering researchers, were amazed at how they wisely recognized, from their close encounters, training and experience, the ups and downs of our proposed actual solutions.
- Educational, and humbling, by how much we had to understand and absorb from the participants, so that we can correct, improve, and refine our solution tools to assure their reliability and effectiveness.

This historical background points out an important learning experience: *Unexpected internal disruptions.* During the workshops, occasionally, certain roles could not be fulfilled as required, because the assigned participants were absent. They were either out of state for other missions, or away for personal reasons.

Although managers tried, and sometime found a work-around, ad hoc backup assignments, the backups had to add and quickly learn new responsibilities on top of their own. They could not possibly be retrained in time for the necessary added, complex procedures under severe time constraints.

Human organizations, large and small, have long addressed and remedied this weakness by a multitude of contingency plans and procedures. Such ad hoc solutions could be fine but limited under uncertain situations and conditions over large-scale safety and security disruptions, requiring rapid response and prevention schemes. For automated procedures, protocols, and cyber-augmented support tools, it presented a major weakness in the otherwise effective system of systems. Those tools, attempting to minimize fumbles and provide the best rapid collaborative actions and smart responses to real threats and disruptions, required a better approach.

We realized that it is necessary to address this weakness by a new collaborative control design principle, which we called Dynamic Lines of Collaborative Command and Control (DLOCCC); or Dynamic Lines of Communication and Control (DLOCC), or in short, Dynamic Lines of Collaboration (DLOC). The learning, adaptive, and evolutionary DLOC we called ELOC. This book explains the details.

For the DLOC research work described in this book, we are grateful to many contributors and collaborators:

Barry K. Partridge, Jim M. Porutalski, Kumares Sinha, Hong Wan, Seokcheon Lee, Elisa Bertino, Juan E. DeBedout, Sigal Berman, Florin G. Filip, Wootae Jeong, Sang Won Yoon, Xin W. Chen, Juan D. Velasquez, Hoo Sang Ko, Manuel A. Scavarda, Hyesung E. Seok, Radhika Bhargava, Rodrigo Reyes Levalle, Lu Zhang, Glenn Candranegara, Luis Restrepo, Arfinandi Ferialdy, Meerant Chokshi, Kwang Hyun Cho, Wonyup Ko, Mohsen Moghaddam, Win P. V. Nguyen, and others, who we may have not listed, but will do so in the next edition. We also wish to thank the friendly Springer staff members, who have helped us throughout the process of preparing and bringing you this book.

Menlo Park, USA   Hao Zhong
West Lafayette, USA   Shimon Y. Nof

Summary

What can we do about disruption events and their cascading effects? Our world in the cyber space expands rapidly with the evolution of intelligent technologies for interaction, communication, sharing, and collaboration. Our systems of machines, software services, and human organizations have become increasingly interdependent, in other words—networked. Disruptions that happen to only a small part of any network tend to escalate. The cyber supports for handling disruptions are often also augmented for the first responders and emergency handlers. That way, they become more responsive and capable of collaborating with each other in controlling the disruptions and preventing their escalation. We are interested in how effectively the collaborations can be supported, and how we can further optimize such support. This book provides the systematic modeling of the complex situations of escalating disruptions and the cycles of dynamic collaborations for the best handling of disruptions. Solution guidelines for optimizing the collaborations are detailed with examples in different application domains: agricultural robotics, civil cyber-physical infrastructure, visual analytics, manufacturing automation, and supply chains. Open-source simulation tools are also released with the book.

Contents

Chapter 1
Introduction

1.1 e-Work, Cyber Physical Systems, and Disruptions: Definitions and Examples

Disruptions occur as interruptions, disturbances, failures, breakdowns, interferences, and other changes in what may be considered a status quo, or normal mode of operations. Our objective in this book is to understand what disruptions are, and how to address them and handle them most effectively.

Of particular interest are disruptions in internetworked e-Work systems, which have emerged since the end of the 20th Century in all modern industry and service systems and systems-of Systems. e-Work (electronic work) is a collection of collaborative, computer-supported, and communication-enabled e-Activities, e-Operations, e-Functions, and e-Support systems that enable other e-Systems and e-Activities in highly distributed organizations of humans and automated machines [49–51, 53, 57, 66, 77].

The current trend of automation includes emerging and upgrading large-scale systems based on integration of intelligent and networked devices, working in an integrated environment. Combining computing, e.g., data storage, processing, mining, cloud services, IIoS (Industrial Internet of Services); and communication capabilities with physical assets, e.g., sensors, robots, IIOT (Industrial Internet of Things), production, infrastructure and service, enables the design and operation of systems that are inter-connected and evolve into Cyber-Physical Systems (CPSs).

New challenges have arisen in the paradigm of e-Work and CPS, because these systems are increasingly integrated, with pervasive connectivity. The motivation for high-scale connectivity is that any CPS consists of a large number of elements which are interdependent on each other through cyber and physical links.

The services required by clients that are served by, or in a CPS, e.g., repairs and preventive maintenance, influence the design and behavior of the network of clients and their service providers (which can themselves be CPSs). The interdependencies

© Springer Nature Switzerland AG 2020
H. Zhong and S. Y. Nof, *Dynamic Lines of Collaboration*,
Automation, Collaboration, & E-Services 6,
https://doi.org/10.1007/978-3-030-34463-4_1

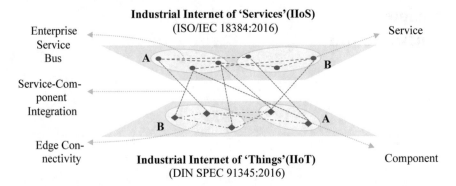

Fig. 1.1 Service-component integration in a cloud manufacturing service for a collaborative network of organizations or systems-of-systems, A and B [39]

also affect the way by which the service providers (suppliers) can deliver timely services to the CPS clients. With the increasing deployment of CPSs, e.g., cyber-physical infrastructure for utilities supply, autonomous connected cars, drones, manufacturing robot teams, besides the interdependent service requests, the services become even more complicated and require the collaboration and matching of multiple servers at the same time (see example in Fig. 1.1).

1.1.1 Disruptions: Types and Definitions

What is a disruption? The simple definition is: A disturbance or an obstacle that interrupt the normal operation, activities, and processes.

While disruptions are often considered to have negative impact, they can also have positive effects: They can lead to improvements and motivate progress. For example, a machine breakdown can cause production and service losses and can cause accidents; it can also motivate new preventive maintenance procedures and safety devices. Car accidents, a major disruption, led to the invention of the seat-belt. Digital cameras replaced film cameras and disrupted the film industry; at the same time enabling progress in the quality and convenience of photography. (Consider the progress in taking digital dental photos, now also integrated as input to automatic manufacture of personalized dental crowns.)

Disruptive innovations, such as digital cameras and the steam engine, impact legacy technology and business, often changing "business as usual" (Table 1.1; Fig. 1.2). They typically emerge from technological uniqueness and clear business model advantages. Disruptive innovations can be historically significant, as urbanization stemming from the industrial revolution, and less dramatic while still disruptive, such as the emergence of bottled water.

Table 1.1 Disruptive innovations features: examples

Innovative disruption	Legacy business/industry	Disrupter	Disrupted vulnerability	Difficult to disrupt
Automobiles	Railroads	Cars and trucks	Schedule, accessibility	Lower cost
Sharing economy	Cars and trucks	Shared service, e.g., electric scooters	Congested traffic and parking	Group travel
Online sales	Car dealerships	Online services	Convenient access	Dealers can also provide it
Electric cars	Legacy car engines	Electric batteries	Cost of fuel and maintenance	Driving range
Autonomous cars	Legacy cars and trucks	Autonomous vehicles	Human drivers	Schedule

Fig. 1.2 Disruptive innovations examples: Collaborative robots serve autonomously in smart warehouses and distribution center, and autonomous cars disrupt transportation. These robots are themselves subject to two types of disruptions: (1) physical disruptions, e.g., infrastructure failures, robot breakdowns, navigation mistakes, and collision disruptions; (2) logical disruptions, e.g., communications, planning errors, inventory errors, and schedule conflicts

Other examples [83] are automobiles that disrupted horse carriages; hotels and airlines that offer less amenities and convenience for lower price; personal computer (PC) innovations that disrupted mainframe and minicomputers by component standardization and major manufacturing cost advantages, while offering continuously improving performance and functions; car sharing replacing car ownership; and online grocers disrupting retail groceries.

Common to all types of disruption is the fact that somebody is impacted negatively. Most innovative disruptions have longer-term impacts. Yet, they have to be handled in a similar manner to the identification, handling, and mitigation of other, common disruptions. The latter include disruptions that are caused by natural disaster and emergencies; security and safety disruptions; and disruptions to infrastructure, equipment, machine, e-Work and CPS disruptions. A summary of typical disruptions and their examples is given in Table 1.2.

Table 1.2 Disruption types and their characteristics

Disruption type	Who is disrupted?	Features	Handling and control	Example reference
Disruptive innovation				
Kiva, a mobile robot	Warehouse and material handling operations	A company with a warehouse for picking and packing e-commerce orders can save workers a lot of running back and forth fetching items from shelves if it had a small crew of robots bringing the shelves to them	Deploy the mobile robots, and train human workers to collaborate with them	[42]
Disruptive technologies and business models	Legacy and current business and practices	Comparison of vulnerabilities and defensible assets	Leveraging/extending core advantages; participating with your own disruptive innovation	[83]
Self-driving car	Traditional cars	Collaboration networks enable coopetition and marketing	Coopetition between networks, organizations, and businesses	[65]
Damaging disruption				
Deviations from plans of resource performance and costs	Project schedule and project management	How to react when an ongoing project is disrupted	Recovery process optimized by hybrid mixed-integer procedure	[96]
Adverse underwater environmental conditions and system constraints, resulting in an intermittently connected, delay/disruption tolerant network	Underwater sensor network for aquatic applications	Simulation analysis of Trade-off among delivery ratio, end-to-end delay, and energy consumption	Specialized adaptive routing protocols for applications having different requirements for different types of messages	[91]

(continued)

Table 1.2 (continued)

Disruption type	Who is disrupted?	Features	Handling and control	Example reference
War, natural disasters	Availability pf critical resources for military logistics	Multi-agent based crisis management supply chain optimization model	Model of subjective variables (risk, uncertainty and vulnerability) and objective variables (inventory levels, delivery times, and financial loss	[30]
Rail track intrusion; medical emergencies; weather extremes; track failures	Rail transit passengers service	Survey of 48 international passenger rail transit agencies about disruption types and their handling	Preplanned policies and backup transportation resources include bus bridging, and transfer to next train, or alternative train route	[59]
Disasters causing large-scale, correlated cascading failures	Telecommunication networks and their populations of users; severely hampered rescue operations; loss/disruption of critical content/services shared over the cloud; security and economic well-being of society	Ensure end-to-content connectivity	Algorithms to manage post-disaster traffic deluge, and failures repair; overcome cascading failures	[43]
Raw material shortages, high prices, supply restrictions	Companies, technologies, and economies that depend on the disrupted raw materials	Analysis of 18 vulnerability assessments of copper and of rare earth neodymium	A set of six vulnerability indicators frequently used and recommended for raw materials assessment decision makers	[26]
Supply disruptions	Vulnerabilities by lean operations and supply network complexities	Based on 180 studies, categorize disruption risks impact evaluation; mitigation strategies	Decisions models for: strategic, sourcing, contracts and incentives, inventory, facility location, disruption evaluation	[73]

(continued)

Table 1.2 (continued)

Disruption type	Who is disrupted?	Features	Handling and control	Example reference
Power systems under terrorist attacks	Power grid and its customers	Identify critical areas which can cause greater damage by cascading failures, and should be given priority of protection	Vulnerability models and metrics enable comparative analyses and development of protection strategies	[80]

1.1.2   Disruptions as Failures in Complex Networks Due to Vulnerabilities

Disruptions fail complex networks of production, supply, and service facilities, enterprises, and infrastructure due to increased vulnerabilities:

- Increase in exposed nodes over longer networks;
- Increase of distance and time;
- Decrease in flexibility;
- Decrease in backup and redundancy;
- Other weaknesses, such as lack of preparedness, mitigation strategies, and insufficient training.

In real-world CPSs, disruptions often have the characteristics of a cascading failure under perturbation. A well-known example is the blackout of power grids which occur from time to time [31]. The significance of this type of disruption is that due to cascading failures, even a single element failure may lead to catastrophic consequence [80].

The analysis and modelling of cascading failure has only been available in recent decades. There are two main schools of thought for modelling cascades in complex networks: the load-based, and the threshold-based. The former, load-based effect, is based on the concept of dynamical redistribution of flow in networks [12]. This model is commonly used to analyze the cascading dynamics in networks where the elements are subjected to loads, e.g., power grid, water distribution networks [34, 70, 71]. The load-based effect is not used by the DLOC model.

The threshold-based effect is modeled considering the diffusion of information or propaganda in social networks, where each individual's state depends on its neighbors [81, 24]. Under the threshold-based model, global cascade or full-size cascade occur according to a "cascade window" which varies according to the degree and threshold value. In the initial model, cascade is triggered by a single failing node, and the rest of the nodes will also fail, if the threshold is met. Further application of this model has resulted in a number of generalizations for different networks. In social contagion (by viruses, information, innovation), a generalized model of has been developed by integrating interdependent interaction models [17]. The model has also been generalized using analytical approach in modular networks [22], degree-correlated networks [16], and networks with adjustable clustering [25]. It has also been found that different failure initiator selection and number would influence the final size of cascade [72].

The threshold-based model is similar to other social contagion models, such as the SIS model, where these models consider the fraction of "failure" neighbors to determine a given node's probability of also becoming faulty [17]. In information diffusion theory, sociologists have long argued that "bridges" between disjoint community clusters promotes the diffusion of information or diseases [23]. This property was further confirmed by the small-world network model [82], where links between otherwise distant nodes are created by rewiring that network model by a regular graph.

It was found that disease infection spreads much easier and quicker in this network. Nevertheless, the type of cascades assumed in the aforementioned studies were *simple propagation*; one "failed" neighbor is sufficient to transmit information or tilt the status of its neighbor into "failed" as well. The other type of cascade is the *complex propagation*, where it requires a minimum threshold of neighbors in "failed" status to enable a given node to adopt the same "failed" status [24].

Both of the above cascade types are addressed in the Watts Global Cascade Threshold model by adjusting the cascade threshold [81]. Complex propagation typically unfolds in clustered networks, or within cluster modules of networks. This phenomenon was exemplified in the studies of recruitment patterns for social movements; they are typically effective in locally dense network of relationships [36, 37].

For *complex propagation*, it was shown that "bridges," or random edges connecting node clusters, can actually inhibit the cascade growth process [6]. In fact, other studies proved that the occurrence of connected clusters in networks are the only obstacles to cascades [19]. Specifically, given the cascade threshold is qq, a failure/disease/information cannot propagate into a different node cluster if the next given node cluster has a clustering density (coefficient) greater than $1 - qq$.

Because of the increasing interest in creating or preventing disruptions of (or through) networks, there have been significant recent advances in the study of cascading failures in complex networks. Now there are opportunities to advance research in this area by the integration of the DLOC model. DLOC and dynamic lines of collaboration to collaboratively prevent disruptions present a new class of research problems in the emerging e-Work systems, cyber-physical systems, and cyber-physical infrastructure. Past findings and research regarding cascade behavior can be applied with DLOC-CDR, Dynamic Lines of Collaboration - Collaborative Disruption Response [93] to aid in developing useful design guidelines for disruption responders in CPI networks.

1.1.3 Disruption Handling by Repair, Mitigation and Control

Extensive work has been reported about disruption handling in terms of resilience to manage the unexpected [11, 69]; the role of decision support and training [10, 54–56, 87–89]; and the role of collaboration in supply networks resilience [62, 63, 67].

Recent work in network disruption mitigation and control can be broadly grouped into two general directions:

- The first direction discusses how to design a robust network by identifying the critical elements of the network to maintain connectivity, and connectivity reliability; maintain single component connected graph topology. This approach can also be implied as pre-disruption mitigation approach.
- The second approach, although not completely exclusive of the first, discusses post-disruption mitigation.

Prior efforts to identify critical elements-connectivity have mainly focused on the application of design of sensor- and radio-communication networks. The minimum number of neighbors needed to maintain connectivity in a random radio network was previously investigated through simulation [48]. They found the "magic number" of minimum neighbors to be three to eight. For wireless sensor networks, the number is estimated by a logarithmic function of the total number of nodes [85]. Subsequent research has extended this model by taking into account various conditions. For example, Dong et al. [18] found the lower bound probability of a wireless sensor network to be connected under Rayleigh Fading as a function of minimum node density. Connectivity properties were also studied in large scale sensor networks as a means to optimize multi-path routing [60].

Another approach for pre-disruption mitigation is by providing redundancy. Chen and Nof [9] and Chen [7] investigated genetic algorithms to be used in modelling low-cost fault tolerant structure of Multi-Enterprise Networks. Inspired by the Fault Tolerance by Teaming principle of CCT [52], a new design paradigm called Resilience by Teaming [61, 62] has been developed and validated on several supply networks to provide better network resiliency under disruptions.

Connectivity reliability analysis has also been advanced on several other types of real-world networks. In transportation networks, research has been conducted to aid post-disaster road network recovery decisions [33]. The road network was modeled as a weighted flow graph, where the flow represents time-varying traffic. By assessing connectivity reliability of different recovery scenarios, an optimal decision can be found to minimize the total travel time cost between each pairs of nodes. Optimal resource allocation (cost) for partial road network recovery has also been investigated using Lagrangian based heuristic algorithm [35].

Still under the category of pre-disruption mitigation strategies, other researchers have investigated network design for self-healing telecommunication networks, by utilizing spare capacity planning. The network in these projects are modeled as bi-directional weighted networks, having multiple commodities between different source and sink nodes (multi-commodity flow problem). There are two main basic methods devised by these researchers: link restoration, and path restoration. Link restoration allocates spare capacity to the links, so that a faulty link's flow can be rerouted through an alternate path, using the spare capacities of the links in the network [78].

Other researchers who also applied the above methods have mainly focused on developing algorithms and heuristics to compute optimal rerouting policy (Krishnamurthy et al. [32]. Path restoration, on the other hand, considers each path disrupted by the link failure separately, and rerouted over an alternate path between the source and sink nodes [29, 44]. This method, although it requires more computation power, results in a more efficient spare capacity planning. Research developments in pre-disruption mitigation strategies have equipped network planners with better insight on the design of a more resilient client network, especially in weighted telecommunication networks.

The N2N challenges that DLOC addresses, however, require the properties of Online Service and Cascading Failures [92, 93]. These properties have not been addressed in the aforementioned works. Post-disruption mitigation strategies research efforts are commonly found in protection of vital infrastructure networks. In Water Distribution Networks (WDN), an emergency model was developed to redistribute water pressure and flow to prevent cascading failures due to overload [71]. Nodes in the network graph represent water reservoirs, consumers, and tanks, while graph edges represent pipes, pumps, and valves. The loads (flows) are assumed to be dynamic, such that it will cause edge failure if it exceeds the flow capacity. In this model, external emergency resource exists to fix failed elements (nodes/edges) and fix requirement signals are triggered by a certain threshold. External resources here are assumed to be unlimited.

Power grids utilize a load shedding strategy to balance overall demands with electricity availability. In this case, the power grid network is abstracted as a graph, where the nodes represent buses (loads and generators), and edges represent electricity lines [1, 3, 84]. Another method is proposed by Motter [40], where "insignificant nodes" that contribute more load to the network than they handle are removed to reduce the size of cascading failures. Additional examples of recent disruption handling cases, their mitigation and control approaches are summarized in Table 1.3.

In summary, the post-disruption mitigation strategies presented offer a robust method to mitigate disruptions, especially for client network control during disruption. They have not yet, however, included the external service network aspect of the DLOC.

1.1.4 Service Resource Allocation and Protection Priority

Service resource allocation problems found in recent work are commonly related to facility location problem, or k-center problem in graph theory. Given a weighted network, this problem is concerned with optimal placement of facilities to minimize transportation cost across the network. There has been significant amount of research

Table 1.3 Disruption handling, mitigation, and control—case examples

Disruption type	Disrupted	Handling features	Mitigation and control	Example reference
Disruption events, e.g., power failure, market change, machine breakdown, supply shortage, worker no-show	Production plans and schedules	Disruption management by solving dynamic programming models of cost and demand disruption cases	Recovery plan to suit the changed environment brought about by the disruption, and seek to be close to the initial plan to not cause too much customer dissatisfaction or inconvenience	[86]
Disruptive events in capacity of production, storage, changeover, lost demand	Manufacturing enterprises	Modeling resilience dynamics and control strategies	Responding to disruptive events by operational and inventory redundancies with a network model of interconnected suppliers, producers, customers and transportation links	[27]
Frequent and rare disruptions: (1) intentional disruptions, e.g., terrorist attacks; (2) unintentional man-made catastrophes, e.g., major accidents; (3) natural catastrophes, e.g., tsunami, hurricane, disease outbreaks	Supply chain practices Globalization De-centralization Outsourcing Single sourcing Just in time Contract litigation	Catastrophe classification framework Cost/benefit tradeoffs and selection of mitigation strategies	Mitigation by Proactive strategies Advance warning strategies Coping strategies	[74]

(continued)

Table 1.3 (continued)

Disruption type	Disrupted	Handling features	Mitigation and control	Example reference
Within a large retailer supply chain: inherent high frequency disruption risks; infrequent disruption risks	Inbound/outbound supply sources information and financial flow systems Product flow, storage, inventories Assets normal activities	Case studies-based lessons learnt for generic mapping of low, medium, and high risk disruption types, and methods to manage/control them	Disruption management by mitigation strategies for: supply, e.g., better coordination; flexible capacity; contingency plans Demand, e.g., multi-sourcing Other risks, e.g., operational cost reduction	[58]
Disruptions that occur in a manufacturing system, e.g., material unavailability, resource failures, operators unavailability, rush orders	Manufacturing systems and facilities; order deliveries	Biological immune system-inspired multi-agent software tools to assist decision makers in dealing with various types of disruptions	Detect, evaluate, react, coordinate by immune system methodology	[15]
Frequent, recurring and regular disruptions: port congestion, variable terminal productivity, unexpected waiting in port channel access Rare and less regular: Bad weather Labor strikes	Shipping vessels service and schedules; port services	Recovery of the affected schedule Is formulated as a multi-stage stochastic control model to minimize the total expected fuel cost and delay penalty	Real time schedule recovery policies, by optimal control policies with strategies addressing different types of disruptions	Li et al. (2016)
Uncertainties about demand for and availability of healthcare services	Patients' waiting time for healthcare	Hierarchical location-allocation model to optimize services under disruption of treatment resources.	Decision support system based on the queuing model for adopting strategies and policies to assure reliable healthcare service	[90]

on this problem. Exact and approximate algorithms have been developed to find the optimal placement [14]. A subset of this problem deals not only with minimizing cost, but also maximizing coverage. The common name for this set of problems is the maximum coverage/shortest path problem [13].

Location and covering problems in undirected and directed flow networks have been studied [75, 76]. The optimal solution of both of these problems can be obtained in polynomial time. Resource allocation (facility location) for post-disaster management also requires the facility to have maximum coverage of the affected area with respect to minimum routing cost [79]. One of the key elements in this work is the resource constraint on the number of resources to be deployed, vs. minimum cost routing, which was solved by integer programming. A sensor location problem in traffic networks has also been investigated to find the minimum number of sensors, such that information on flow volume in a specific path can be obtained [21].

The maximal coverage/shortest path problem aligns with resource allocation problems applied in DLOC, where a disruption handling server has to be initially positioned in nodes that minimize overall expected routing cost, as well as providing maximum coverage of the network, in terms of group allocation. The approach employed in this problem, however, cannot be directly adopted to the collaborative disruption response problem in DLOC. As stated in Zhong and Nof [93], Zhong [92] the positioning of server agents in client networks also functions as error prevention in the supervised nodes/links. Thus, there is another objective of maximally positioning the server agents to protect important (critical) nodes.

Due to cascade of overload failures, the highly heterogeneous distribution of *loads* of real-world network makes them vulnerable to attacks, such that an avalanche of failure nodes (cascading) may occur by disabling a single (or several) key nodes [41]. A better protection strategy of a client network can be developed by also taking into account this fact, i.e., priority protection of high priority, vulnerable nodes. Several researchers have investigated network survivability under the failure of this set of high priority nodes and identify them for priority protection. Cruciti et al. [12] showed that in weighted networks, the vulnerable nodes are the ones with the largest load. In a fiber infrastructure network, a polynomial time algorithm has been developed to simulate several node failures (single or set) to identify the vulnerable nodes [45]. Vulnerable nodes in a directed water distribution network were identified by assessing the ratio between discrepancy of failures cascade, and direct failure and total number of initial node [71].

The previously mentioned works have investigated node vulnerability under the consideration of cascading overload failures, typically in flow networks. Notwithstanding the importance of flow continuity in networks, vulnerable nodes can also be identified by analyzing the network topological structure. Graph (network) centrality measures have been developed, including degree centrality, and betweenness centrality [4, 20, 46]. These previous centrality measures are dominated by the elements' degree. A newly developed measure, named bridging centrality measure, aims to identify the most important component in the networks by exploiting graph properties of cut edges/vertices and clustering [28]. Cut vertices/edges denote the elements of a graph (network) where if removed would increase the number of connected components. This measure is essential in real-world networks which mainly exhibit a modularity structure [47].

In summary, recent research advanced service resource allocation, and network component protection priority with significant contribution in understanding the failure dynamics of disruptions, and protection of complex network. Nevertheless, these two areas are still missing the necessary approach of collaborative disruption response. At the same time, these findings contribute important advances to be integrated with the improved strategy of collaborative disruption response, as explained later in the book. Thus, it is possible to improve the service allocation protocol in DLOC by taking into account both the optimal service allocation, and a differentiated network component protection priority approach.

1.1.5 Four Examples of Disruptions in Connected e-Work

The servers and clients in the following examples are all different, but they share the similar structure of providing services by a group of servers ("suppliers") to a group of clients ("customers").

1. *Collaborative Disruption Response (CDR).* Failures in Cyber-Physical Infrastructures (CPIs) and supply networks can propagate to large-scale catastrophes due to the interdependency of system components [68, 94, 95]. A team of disruption responders can optimize their teamwork deployment by collaboratively preventing, detecting and recovering from cascading failures (see Fig. 1.3).
2. *Dynamic teamwork for manufacturing design and assembly.* The design and assembly of engineered products often requires the teamwork of multiple designers, suppliers, operators, assemblers, and maintenance and repair personnel, each supplying diverse domain knowledge (see Fig. 1.4). For instance, in furniture manufacturing [5] design and operational conflicts and human/machine errors on the production line can propagate and result in faulty, lower quality furniture products.
3. *Reconfigurable End-Effectors* (REEs) for automated harvesting. Multiple targets require interdependent harvesting operations, because the operation for one target (e.g., pruning) can affect operations for other targets [2, 94]. To design a cost-effective REE configuration means to optimize the collaborations of robot manipulator gripper parts and tools with a predictive planning of activities and grasped or handles objects for better productivity and grasp quality (see examples in Table 1.4).
4. *Distributed denial of service* (DDoS). A negative example of collaboration among attackers can be found in the distributed denial of service attack on Internet infrastructure [38] and on smart grids [64]. At the same time, collaborative detection and filtering have been developed to counter those attacks [8]. Distributed bots collaborate to disrupt by attacking distributed services of domain name servers, routing protocols, and other networked system components. This problem can

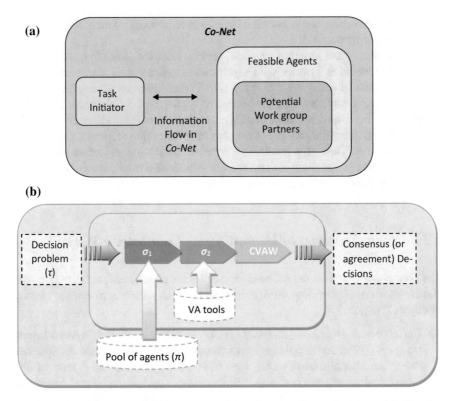

Fig. 1.3 Collaborative disruption response (CDR): **a** responder team formation and allocation (Co-Net: collaborative networked team-members); **b** co-Insights Framework for the response team collaborative decisions and actions (VA: visual analytics; $\sigma 1$: task–participant matching mechanism; $\sigma 2$: Participant–task matching mechanism; CVAW: collaborative visual analytics workspace; HUB-CI: virtual cloud hub for collaborative intelligence among team participants)

only be solved with coordinated actions of multiple Internet or grid participants and service providers. Distributed denial of service disruption can cause severe propagation of disruptions throughout the system-of-systems shown in Fig. 1.1.

The common critical challenge in the above four examples is how to design and organize the interacting teams and their disruption handling services for effective and efficient, timely response to disruptions. Furthermore, preventing as many as possible disruptive potential errors and conflicts can reduce costs, damages, and delays. It can also eliminate irreparable and irreversible damages.

On one hand, teams should include the highly skilled agents (designers, operators, bots, and other human and cyber entities) who can provide the most meaningful skills and insights. On the other hand, agents (workers) must include novice human and cyber agents for training, knowledge transfer, machine learning, and backup.

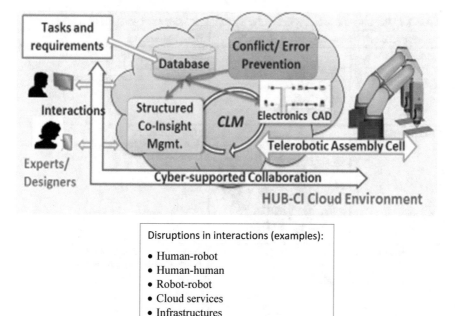

Fig. 1.4 Teams of cyber and human agents and potential disruptions (CAD: computer aided design; CLM: collaborative lifecycle management; HUB-CI: virtual hub of collaborative intelligence shared among the teams) [93]

Experience shows that with the correctly designed and implemented disruption response systems, services and protocols, including shared resources (e.g., inspection, testing and repair resources), significantly better product/service quality at lower cost can be achieved. The above background examples and their mappings to services are summarized in Table 1.5.

1.2 The Network-to-Network (N2N) View of Collaborative e-Work

The elements of collaborative e-Work form a complex structure of collaborative interaction dynamics. By characterizing the structure, it enables the fundamental understanding of how to design and control the collaboration.

Following the examples in Sect. 1.1, two networks can be identified:

- Client network: This network represents the supply, production, service, and infrastructure installations, systems, systems-of-systems. In other words, the target network (target of disruptions).

Table 1.4 Reconfigurable end-effectors for harvesting missions and potential disruptions

Function at time t	Reconfigured end-effector	Disruption examples
Pick small fruit		Fruit size incompatible Fruit falls away Damage to end-effector from spoiled fruits
Pick large fruit		Mixed fruit sizes Branches prevent large end-effector access to fruits Unpredicted fruit size
Prune bypass, branch		Pruned branches damage end-effector Pruned stems get entangled in end-effector

Table 1.5 Examples of disruptions in networked e-Work

Networked e-Work	Clients	Servers	Disruptions service challenge
Cyber-physical infrastructures	Smart grid, water distribution, air transportation, etc.	Responders, repair-agents	Cascading failures
Reconfigurable manufacturing & harvesting	Products that requires specialized handling, fruits, etc.	Reconfigurable peripherals and robot parts	Costly reconfiguration and delays
Manufacturing design & assembly	Tasks of design, assembly, etc. that require diverse domain expertise.	Teams of human and cyber agents (designers and operators)	Inefficient insight/skill sharing and tacit knowledge sharing
Information and communications system and services	Distributed services of domain name servers, system components	Distributed bots	Distributed denial of service to Internet infrastructure

Fig. 1.5 The interactions in network-to-network services of e-Work

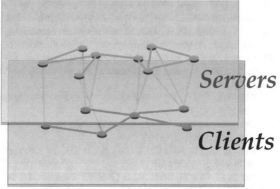

Servers

Clients

———— Interdependency in the client network

———— Interaction between the client network and the service team

———— Collaboration in the service team

- Server network: The network of agents responsible for disruption handling and control. This network includes the repair, reconstruction, maintenance, and recovery agents.

An abstract representation of N2N services is illustrated in Fig. 1.5. While the clients and servers are grouped into two networks, the challenge becomes how to provide effective Network-to-Network (N2N) services in these complex e-Work systems.

References

1. Aponte EE, Nelson JK (2006) Time optimal load shedding for distributed power systems. IEEE Trans Power Syst 21(1):269–277
2. Berman S, Nof SY (2011) Collaborative control theory for robotic systems with reconfigurable end-effectors. In: Proceedings of 21st international conference on production research, Stuttgart, Germany
3. Bevrani H, Tikdari AG, Hiyama T (2010) An intelligent based power system load shedding design using voltage and frequency information. In: Proceedings of the modelling, identification and control (ICMIC) international conference, pp 545–549
4. Borgatti SP (2005) Centrality and network flow. Soc Networks 27(1):55–71
5. Candranegara G, Zhong H, Nof SY (2015) Conflict & error mgmt based on collaborative control theory: a case study in the furniture industry. In: Proceedings of the 23rd international conference on production research
6. Centola D, Eguiluz VM, Macy MW (2007) Cascade dynamics of complex propagation. Physica A 449–456
7. Chen J (2002) Modelling and analysis of coordination for multi-enterprise networks. Doctoral dissertation, School of Industrial Engineering, Purdue University

8. Chen Y, Hwang K (2006) Collaborative detection and filtering of shrew DDoS attacks using spectral analysis. J Parallel Distrib Comput 66(9):1137–1151
9. Chen J, Nof SY (2000) Multi-enterprise networking. In: Proceedings of international conference on manufacturing systems: innovations for the 21st century. Ann Arbor, MI
10. Chen X, Nof SY, Partridge BK, Varkonyi I, Nakanishi YJ (2006) Security awareness and alertness training in states departments of transportation. In: Proceedings of TRB-85, Washington, DC, Jan 2006
11. Chopra S, Sodhi M (2014) Reducing the risks of supply chain disruptions. MIT Sloan Manage Rev 55(3):73–80
12. Crucitti P, Latora V, Marchiori M (2004) Model for cascading failures in complex networks. Phys Rev E 69(4):045104
13. Current JR, Re Velle CS, Cohon JL (1985) The maximum covering/shortest path problem: a multi-objective network design and routing formulation. Eur J Oper Res 21(2):189–199
14. Current J, Min H, Schilling D (1990) Multi-objective analysis of facility location decisions. Eur J Oper Res 49(3):295–307
15. Darmoul S, Pierreval H, Hajri-Gabouj S (2013) Handling disruptions in manufacturing systems: an immune perspective. Eng Appl Artif Intell 26:110–121
16. Dodds PS, Payne JL (2009) Analysis of a threshold model of social contagion on degree-correlated networks. Phys Rev E 79(6):066115
17. Dodds P, Watts DJ (2004) Universal behavior in a generalized model of contagion. Phys Rev Lett 92(21):218701
18. Dong J, Chen Q, Niu Z (2007) Random graph theory based connectivity analysis in wireless sensor networks with Rayleigh fading channels. In: 2007 Asia Pacific conference on communications, Bangkok, p 123
19. Easley D, Kleinberg J (2010) Cascading behavior in networks. Cambridge University Press, Cambridge
20. Freeman LC (1977) A set of measures of centrality based on betweenness. Sociometry 40:35–41
21. Gentili M, Mirchandani PB (2005) Locating active sensors on traffic networks. Ann Oper Res 136(1):229–257
22. Gleeson JP (2008) Cascades on correlated and modular random networks. Phys Rev E 77(4):046117
23. Granovetter MS (1973) The strength of weak ties. Am J Sociol 78(6):1360–1380
24. Granovetter M (1978) Threshold models of collective behavior. Am J Sociol 83(6):1420–1443
25. Hackett A, Melnik S, Gleeson J (2011) Cascades on a class of clustered random networks. Phys Rev E 83(5):056107
26. Helbig C, Wietschel L, Thorenz A, Tuma A (2016) How to evaluate raw material vulnerability— an overview. Resour Policy 48:13–24
27. Hu Y, Li J, Holloway LE (2008) Towards modeling of resilience dynamics in manufacturing enterprises: literature review and problem formulation. In: Proceedings of the 4th IEEE conference on automation science and engineering, Washington DC, USA, pp 279–284
28. Hwang W, Choe Y-R, Zhang A, Ramanatha M (2006) Bridging centrality: identifying bridging nodes, scale-free networks
29. Iraschko RR, Grover WD (2000) A highly efficient path-restoration protocol for management of optical network transport integrity. IEEE J Sel Areas Commun 18(5):779–794
30. Kaddoussi A, Zoghlami N, Zgaya H, Hammadi S, Bretaudeau F (2011) Disruption management optimization for military logistics. IFIP Adv Inform Commun Technol 364:61–66
31. Kadloor S, Santhi N (2010) Understanding cascading failures in power grids. Los Alamos National Laboratory
32. Krisnamurthy S, Chandrasekaran R, Venkatesan S, Dawande M (2003) Highly efficient spare capacity planning for generalized link restoration. In: Proceedings of 12th international conference on computer communications and networks, pp 47–52
33. Li B, Hu X, Xie B (2009) Transportation network reconstruction for natural disasters in the emergency phase based on connectivity reliability. In: Proceedings of the 2nd international conference on transportation engineering, pp 2963–2968

34. Li S, Li L, Jia Y, Liu X, Yang Y (2013) Identifying vulnerable nodes of complex networks in cascading failures induced by node-based attacks. In: Mathematical problems in engineering
35. Liu L, Qi X (2014) Network disruption recovery for multiple pairs of shortest paths. In: 2014 Proceedings of the 11th International Conference on Service Systems and Service Management (ICSSSM), Beijing, China
36. McAdam D (1986) Recruitment to high-risk activism: the case of freedom summer. Am J Sociol 92(1):64–90
37. McAdam D, Paulsen R (1993) Specifying the relationship between social ties and activism. Am J Sociol 99(3):640–667
38. Mirkovic J, Reiher P (2004) A taxonomy of DDoS attack and DDoS defense mechanisms. ACM SIGCOMM Comput Commun Rev 34(2):39–53
39. Moghaddam M, Nof SY (2017) Best matching theory & applications. Springer ACES Series (Automation, Collaboration, and E-Services), Berlin
40. Motter AE (2004) Cascade control and defense in complex networks. Phys Rev Lett 93
41. Motter AE, Lai Y (2002) Cascade-based attacks on complex networks. Phys Rev E 66:065102(R)
42. Mountz M (2012) Kiva the disrupter [mobile robot]. Harvard Bus Rev 90(12):74–80
43. Mukherjee B, Habib M, Dikbiyik F (2014) Network adaptability from disaster disruptions and cascading failures. IEEE Commun Mag 52(5):230–238
44. Murakami K, Kim HS (1998) Optimal capacity and flow assignment for self-healing ATM networks based on line and end-to-end restoration. IEEE/ACM Trans Networking (TON) 6(2):207–221
45. Neumayer S, Zussman G, Cohen R, Modiano E (2011) Assessing the Vulnerability of the fiber infrastructure to disasters. IEEE/ACM Trans Networking 19(6):1610–1623
46. Newman ME (2001) Scientific collaboration networks. II. Shortest paths, weighted networks, and centrality. Phys Rev E 64(1):8577–8582
47. Newman ME (2006) Modularity and community structure in networks. PNAS 103(23)
48. Ni J, Chandler S (1994) Connectivity properties of a random radio network. IEE Proc Commun 141(4):289–296
49. Nof SY (2003) Design of effective e-work: review of models, tools, and emerging challenges. Prod Plan Control 14(8):681–703
50. Nof SY (2004) Transportation security awareness training. In: Proceedings of road school, Purdue University, Mar 2004
51. Nof SY (2006) Collaborative e-work and e-manufacturing: challenges for production and logistics managers. J Intell Manuf 17(6):689–701
52. Nof SY (2007) Collaborative control theory for e-work, e-production, and e-service. Ann Rev Control 21(2):281–292
53. Nof SY, Silva JR (2018) Perspectives on manufacturing automation under the digital and cyber convergence (invited). Polytechnica 1:36–47
54. Nof SY, Yoon SW (2005b) MDI: a transportation security mock drill for Indiana. In: Proceedings of Indiana Road School, Mar 2005
55. Nof SY, Velasquez JD, Chen X, Jeong W, Restrepo L (2004) Identification, design and delivery of awareness training. In: Proceedings of Indiana Road School, Mar 2004
56. Nof SY, Velasquez JD, Chen X, Jeong W, Yoon SW (2005) Identification, design and delivery of awareness training. In: Proceedings of Indiana Road School, Mar 2005
57. Nof SY, Ceroni J, Jeong W, Moghaddam M (2015) Design with collaborative control theory. In: Revolutionizing collaboration through e-work, e-business, and e-service. Springer, Berlin, pp 33–75
58. Oke A, Gopalakrishnan M (2009) Managing disruptions in supply chains: a case study of a retail supply chain. Int J Prod Econ 118:168–174
59. Pender B, Currie G, Delbosc A, Shiwakoti N (2012) Planning for the unplanned: an international review of current approaches to service disruption management of railways. In: Proceedings of the 35th Australasian Transport Research Forum, ATRF

60. Pishro-Nik H, Chan K, Fekri F (2009) Connectivity properties of large-scale sensor networks. Wireless Netw 15(7):945–964
61. Reyes Levalle R (2018) Resilience by teaming in supply chains and networks. Springer ACES Series (Automation, Collaboration, and E-Services), Berlin
62. Reyes Levalle R, Nof SY (2015) Resilience by teaming in supply network formation and re-configuration. Int J Prod Econ 160:80–93
63. Reyes Levalle R, Nof SY (2017) Resilience in supply networks: definition, dimensions, and levels. Ann Rev Control 43:224–236
64. Roy S (2017) Denial of service attack on protocols for smart grid communications. Secur Solutions Appl Cryptogr Smart Grid Commun 50–67
65. Rusko R, Alatalo L, Hänninen J, Riipi J, Salmela V, Vanha J (2018) Technological disruption as a driving force for coopetition: the case of the self-driving car industry. Int J Innov Digital Econ 9(1):35–50
66. Scheibe KP, Blackhurst J (2018) Supply chain disruption propagation: a systemic risk and normal accident theory perspective. Int J Prod Res 56(1–2):43–59
67. Scholten K, Schilder S (2015) The role of collaboration in supply chain resilience. Supply Chain Manage 20(4):471–484
68. Seok H, Kim K, Nof SY (2016) Intelligent contingent multi-sourcing model for resilient supply networks. Expert Syst Appl 51:107–119
69. Sheffi Y (2015) The Power of resilience: how the best companies manage the unexpected. MIT Press, Cambridge
70. Shuang Q et al (2015) A cascade-based emergency model for water distribution network. In: Mathematical problems in engineering
71. Shuang Q, Zhang M, Yuan Y (2014) Node vulnerability of water distribution networks under cascading failures. Reliab Eng Syst Safety 124:132–141
72. Singh P, Sreenivasan S, Szymanski BK, Korniss G (2013) Threshold-limited spreading in social networks with multiple initiators. Sci Rep 3:2330
73. Snyder LV, Atan Z, Peng P, Rong Y, Schmitt AJ, Sinsoysal B (2016) OR/MS models for supply chain disruptions: a review. IIE Trans 48(2):89–109
74. Stecke KE, Kumar S (2009) Sources of supply chain disruptions, factors that breed vulnerability, and mitigating strategies. J Market Channels 16:193–226
75. Tamura H, Sengoku M, Shinoda S (1990) Location problems undirected flow networks. Trans IEICE 73(12)
76. Tamura H, Sengoku M, Shinoda S, Abe T (1992) Some covering problems in location theory on flow networks. IECE Trans Fundamentals 75(6)
77. Valckenaers P, Van Brussel H (2015) Design for the unexpected: from Holonic manufacturing systems towards a humane mechatronics society. Butterworth-Heinemann, Oxford
78. Veerasamy J, Venkatesan S, Shah JC (1995) Spare capacity assignment in telecom networks using path restoration. In: Proceedings of the third international workshop on modeling, analysis, and simulation of computer and telecommunication systems (MASCOTS'95), pp 370–374
79. Viswananth K, Peeta S (2002) The multicommodity maximal covering network design problem. In: IEEE 5th international conference on intelligent transportation systems, Singapore
80. Wang S, Zhang J, Zhao M, Min X (2017) Vulnerability analysis and critical areas identification of the power systems under terrorist attacks. Physica A 473:156–165
81. Watts D (2002) A simple model of global cascades on random networks. Proc Nat Acad Sci USA 99(9):5766–5771
82. Watts DJ, Strogatz SH (1998) Collective dynamics of small-world networks. Nature 393(6684):440–442
83. Wessel M, Christensen CM (2012) Surviving disruption. Harvard Business Review 90(12):56–64
84. Xu D, Girgis AA (2001) Optimal load shedding strategy in power systems with distributed generation. Power Engineering Society Winter Meeting, pp 788–793

85. Xue F, Kumar PR (2004) The number of neighbors needed for connectivity of wireless networks. Wireless Netw 10(2):169–181
86. Yang J, Qi X, Yu G (2005) Disruption management in production planning. Naval Res Logistics 52:420–442
87. Yoon SW, Velasquez JD, Partridge BK, Nof SY (2008) Transportation security decision support system for emergency response: a training prototype. Decis Support Syst 46(1):139–148
88. Yoon SW, Velasquez JD, Ko HS, Chen X, Nof SY (2009) Collaborative distributed-training control system for transportation and emergency response. In: Proceedings of IIE annual research conference, Miami, FL, May 2009
89. Yoon SW, Velasquez JD, Ko HS, Chen X, Nof SY (2010) Collaborative distributed-training control system for transportation and emergency response. In: Proceedings of IIE annual conference and expo, Cancun, Mexico, May 2010
90. Zarrinpoor N, Fallahnezhad MS, Pishvaee MS (2017) Design of a reliable hierarchical location-allocation model under disruptions for health service networks: a two-stage robust approach. Comput Ind Eng 109:130–150
91. Zheng G, Colombo G, Wang B, Jun-Hong C, Maggiorini D, Rossi GP (2008) Adaptive routing in underwater delay/disruption tolerant sensor networks. In: Proceedings of the IEEE fifth annual conference on wireless on demand network systems and services, pp 31–39
92. Zhong H (2016). Dynamic lines of collaboration in e-work systems. Unpublished doctoral dissertation, West Lafayette: School of Industrial Engineering, Purdue University
93. Zhong H, Nof SY (2015) The dynamic lines of collaboration model: collaborative disruption response in cyber–physical systems. Comput Ind Eng 87:370–382
94. Zhong H, Nof SY, Ozsoy E (2015) Co-Viz: matching tools in collaborative visual analytics. In: Proceedings of industrial and systems engineering research conference, Nashville, USA
95. Zhong H, Ozsoy E, Nof SY (2016) Co-insights framework for collaborative decision support and tacit knowledge transfer. Expert Syst Appl 45(3):85–96
96. Zhu G, Bard JF, Yu G (2005) Disruption management for resource-constrained project scheduling. J Oper Res Soc 56:365–381

Chapter 2
Collaborative e-Work and Collaborative Control Theory for Disruption Handling and Control

2.1 Motivation for Collaborative Handling of Disruptions

When we are disrupted and fall, and somebody offers us a hand to get up, we appreciate it. We are thankful that we do not have to handle the disruption consequences alone; not the misery of our fall, nor the hurt emotions. We also know that with that collaborating hand, there is a high likelihood that our disruption can be handled faster and better. Indeed, approaches to handle and mitigate disruptions with some form of collaboration have been developed and implemented successfully for various types of disruptions. Table 2.1 summarizes several such examples.

2.2 Overview of Collaborative Control Theory

As production and service systems become increasingly complex, the number of components and the interactions among them increases drastically. Collaboration in such highly interconnected environments becomes a necessity for the achievement of reliable, timely, and cost-effective goals.

Optimizing the collaboration between systems and/or system components, however, becomes ever more challenging than before. Collaborative Control Theory (CCT) is a framework of principles for the design and control of complex systems with multiple collaborating agents in networked organizations and facilities [21].

Collaboration implies the sharing of information, resources, and tasks. CCT studies the design and control of collaboration in e-Work [19, 22]. CCT framework offers a useful direction for optimizing the design and control of emerging networked systems. CCT is involved in each phase of collaboration, and collectively forms collaboration lifecycle management (CLM), as shown in Fig. 2.1.

The collaboration lifecycle can be described as follows.

© Springer Nature Switzerland AG 2020
H. Zhong and S. Y. Nof, *Dynamic Lines of Collaboration*,
Automation, Collaboration, & E-Services 6,
https://doi.org/10.1007/978-3-030-34463-4_2

Table 2.1 Collaborative handling of disruption—examples

Disruption case	Vulnerability disrupted	Features	Collaborative handling	Example reference
Disasters				
Hurricane	Power supply; National security	Speeding restoration by private teams	Collaborating responder teams	[20]
Climate change disaster	Ports and supply chains	Collaboration between research and practice communities	Collaborative adaptation planning process	[1]
Natural disasters	Transportation systems	Graph theory studies of vulnerability and recovery	Cross-disciplinary collaborations to transform knowledge to resilience strategies	[18]
Earthquake	Economic business continuity and livelihood of residents	Multi agent simulation	Stakeholders collaboration by information sharing	[16]
Typhoon	Humanitarian needs	Decision- and goal-oriented approach	Ad hoc collaboration among a variety of actors and organizations	[33]
Urban infrastructure disasters	Transportation, water, and power services	Game theory model of equilibrium in a multilayer infrastructure network	Systematic evaluation of mutual influence between infrastructure and communities	[17]
Emergencies				
Air transportation disruptions	Airport operations	Decision support system to train first responders	Collaborative modeling, simulation, and visualization framework	[11]
Emergency responders	Emergency responder application/task overload	Tool matching and agent based collaborative visual analytics	Collaborative control of workflow and decision making in collaborative visualization	[41]

(continued)

Table 2.1 (continued)

Disruption case	Vulnerability disrupted	Features	Collaborative handling	Example reference
Cyber physical system; smart grid disruptions	Continuous functionality	HUB-CI (high performance computing hub with collaborative intelligence tools)	Dynamic lines of collaboration protocols	[39, 40]
Cyber physical systems disruptions	Large scale disasters due to propagating failures	emergency services and repair operations of cascading failures	Collaborative disruption response with centrality based agent team allocation	[38]
Aerospace industry disruptive innovations	Supply chain management	Value networks; insights from industry	Cooperative competition and complementarity framework	[23]
Urban infrastructure cascading failures	Disruptions cascade within and across gas, water, power, and transportation systems	Game-theoretical equilibrium model of the mutual influence between the infrastructure and communities	Redistribution of supplies critical to life-support	[17]
Production/supply/service network shortages				
Healthcare supply chain disruptions	Hospital supply chain	Causal loop diagrams	Collaborative information sharing & supply chain management	[9]
Supply disruptions	Supply networks	Contingent multi-sourcing protocol	Contingent collaborative suppliers coalition	[30]
Supplier induced disruptions	Supply conflicts in production	Fuzzy-set qualitative comparative analysis	Collaborative responses in buyer-supplier interactions	[28]
Demand disruptions	Supply chain	Information management strategies	Collaborative planning, forecasting and replenishment	[36]

(continued)

Table 2.1 (continued)

Disruption case	Vulnerability disrupted	Features	Collaborative handling	Example reference
Flexible job shop execution disruptions	Schedule execution	Ant colony optimization of multi agent scheduler	Disrupted execution rescheduling by multi agent teams	[37]
Airline operations control disruptions	Scheduled flights and services	Multi agent system analysis and control	Implicit and explicit machine learning and collaborative decision making	[4]
Computer/communications security				
Cyber crime	Critical infrastructure	Cyber and Network security	Intelligent control and collaboration	[27]
Mobile ad hoc networks disruptions	Multiple attackers synchronize their actions to disrupt a target network	Nodes mutually monitor other nodes to identify malicious behaviors	Detection, prevention, and blocking of malicious collaborative attacks	[14]
Distributed Denial of Service (DDoS), and Flash events	Internet based services, financial losses	Survey of methods to detect and prevent such attacks	Collaborative solutions to prevent collaborative attacks	[2]
Telecommunications attacks by selfish and malicious nodes	Wireless data equipment, communications, and delivery	Detection and defense systems, and incentive mechanisms	Human-centric cooperation network protocols	[12]

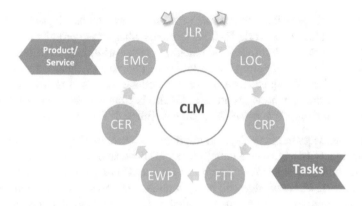

Fig. 2.1 Collaboration Lifecycle Management. Adapted from Zhong et al. [41]

- It begins at the JLR (Join/Leave/Remain) node in Fig. 2.1, where potential participants are mapped into the collaboration teams guided by the JLR
  principles [8].
- The immediate step after forming the collaboration team is to set up the structure
  inside the team. The Lines of Collaboration and Communication (LOC) methods with collaborative visualization techniques are used to map various participants to different interfaces by their preferences, and to build multiple lines of
  accountability, including backups during emergent situations [35].
- When the product requirements arrive at the design team, Collaboration Requirement Planning (CRP) is used to allocate resources to meet the design requirements
  and provide knowledge by searching in knowledge bases [25, 26, 41]. The concrete
  collaboration work begins after CRP.
- Fault-Tolerance by Teaming (FTT) and e-Work Parallelism (EWP) reveal two
  aspects of collaborative e-Work: specialization and cooperation. Different experts
  have different expertise.
- EWP empowers each individual to work with tools in parallel, so each part of the
  collaborative work is handled by the right person or an expert agent [5]. EWP
  derives the optimal degree of parallelism of different components, performing
  different tasks in a system.
- FTT indicates that in general, with intelligent protocols, the teamwork of the agents
  can be resilient to faults and errors even if some of the agents are faulty [15, 29]. The
  set of tools provided by CLM systems can be controlled in both an individual mode
  and a collaborative mode, to realize EWP and FTT. The collaborative network of
  humans and agents, however, is prone to errors and conflicts.
- Conflict and Error Resolution (CER) principle aims to detect errors as early as
  possible to prevent cascading failures in both cyber and physical systems [7].
- The last CLM step applies e-Measurement and e-Criteria (EMC), a set of e-enabled
  evaluation mechanisms and systems developed to examine e-Work effectiveness

[10, 13], and apply them for ongoing evaluation, as well as adaptive and evolutionary learning. In e-Work, EMC determines whether the product or service is ready for testing and delivery, by reaching consensus among experts and product requirements. If a product or service does not meet predefined requirements, or any additional requirements arrive, the collaboration will move into another iteration of the collaboration process. Otherwise, the collaboration team will be restructured (using JLR principles) to a new lifecycle for other projects.

The dynamic teaming mechanism improves the sustainability of product design. When an agent (a designer or an intelligent tool) flows into another team, the knowledge it has will contribute to a different set of agents. Consequently, it can be observed that an increase of knowledge will be shared in the collaboration environment. The next three subsections present three detailed principles that are highly related to the DLOC work.

2.3   Collaboration Requirement Planning

Collaboration Requirement Planning (CRP) enables efficient design and control of automation based on given tasks and available resources. CRP has been applied to handle the planning and optimization of tasks and actions by cooperating robots in complicated tasks [25]. The CRP framework has been adapted to facilitate the design and evaluation of automation systems with reconfigurable end-effectors (REEs, [3]).

Extended CRP with grasp support generates plans for the end-effector selection and the grasp strategies, considering grasp requirements. CRP has two phases: CRP-I is used for work plan generation based on objectives and available recourses; and CRP-II executes in real time for adapting the plan to changing spatial and temporal conditions. For extended CRP with REE, CRP-I determines the configurations of REE based on the constraint of minimal precedence tasks. CRP-II enables dynamic decisions on task assignments to robots in the cell based on their changing availability.

As many components in the automated system are working in parallel, and the tasks are interdependent, the planning of executions and configurations becomes even more challenging. A new framework is required to fulfill the new challenges in N2N servers.

2.4   Conflict and Error Prevention by CTT

Conflicts and errors are unavoidable in a large-scale collaboration of multiple participants. One important advantage of e-Work is that it enables effective methods to cope with errors and conflicts quickly and less expensively. While conventional error detection slows the performance of the design system, using networks of agents operating in parallel, costs can be reduced [6].

The CER principle of CCT has proven that network-aware conflict and error prevention and detection algorithms can efficiently reduce damage and the prevention and detection time, and increase the coverage and preventability of conflicts and errors [7]. Hence, centralized algorithms for centralized networks and decentralized algorithms for decentralized networks improve prevention and detection.

To support the services over an interdependent network, the conflict and error detection should be linked with the recovery of failures caused by the conflicts and errors. If the current conflicts and errors are not detected and recovered, future conflicts and errors may not be prevented, due to the cascading effects. In this book, the optimal schedule of conflict and error prevention, detection, and recovery is explored.

2.5   Fault Tolerance by Collaboration

The structure of a system describes the relations among system components. To determine a team structure, i.e., the allocation of tasks, responsibilities and authority is a fundamental problem in organizational science [31]. Dynamic teams (e.g., temporary alliances, changing virtual organizations) are essential in e-collaboration especially under emergent situations [35].

The lines of collaboration (LOC) are constantly updating inside and between teams. First, it is obvious that different teams need to be formed to satisfy the various task requirements demanded. Moreover, changing team structure is critical for the sustainability of the entire organization [34].

Schools of fish serve as an analogy for human systems. A regrouping of fish schools takes place, as a rule, after each task and disbanding [24]. Hence, quite an intensive exchange of specimens goes on between and among schools; new schools with a new composition are formed after each cycle. The fish passing from one school to another bring with them their own reflexes (knowledge and expertise of how to defend against predators), adding to the "wealth" of other schools.

For human experts, dynamic team formation for decision making provides an effective channel to retain and to transfer tacit knowledge. The notion of team configuration based on functionality, knowledge and expertise is applied in the current research. Dynamic team structure also applies to artificial systems. Autonomous agents establish dynamic teams to improve data and patterns sharing.

In collaborative robotics, dynamic role assignment improves the performance of the overall team [32]. How to determine system configuration quantitatively over time, however, remains an open question. The current work extends the LOC principle to explore how to design the configuration of the service team that enables dynamic lines of collaboration to improve the N2N service performance.

References

1. Becker A, Ng AKY, McEvoy D, Mullett J (2018) Implications of climate change for shipping: ports and supply chains. Wiley Interdiscip Rev Climate Change 9(2) (2018)
2. Behal S, Krishan K, Sachdeva M (2017) Characterizing DDoS attacks and flash events: review, research gaps and future directions. Comput Sci Rev 25:101–114
3. Berman S, Nof SY (2011) Collaborative control theory for robotic systems with reconfigurable end-effectors. In: Proceedings of 21st international conference on production research, Stuttgart, Germany
4. Castro AJM, Rocha AP (2017) Managing disruptions with a multi-agent system for airline operations control. In: Proceedings of advances in practical applications of cyber-physical multi-agent systems conference, Porto, Portugal, pp 307–310
5. Ceroni JA, Nof SY (2002) A workflow model based on parallelism for distributed organizations. J Intell Manuf 13(6):439–461
6. Chen XW, Nof SY (2010) A decentralized conflict and error detection and prediction model. Int J Prod Res 48(16):4829–4843
7. Chen XW, Nof SY (2012) Conflict and error prevention and detection in complex networks. Automatica 48(5):770–778
8. Chituc CM, Nof SY (2007) The join/leave/remain (JLR) decision in collaborative networked organizations. Comput Ind Eng 53(1):173–195
9. Gonul KC, Nowicki DR, Sauser B, Randall WS (2018) Impact of cloud-based information sharing on hospital supply chain performance: a system dynamics framework. Int J Prod Econ 195:168–185
10. Huang CY, Nof SY (2002) Evaluation of agent-based manufacturing systems based on a parallel simulator. Comput Ind Eng 43(3):529–552
11. Jafer S (2014) Collaborative modeling, simulation, and visualization framework for airport emergency. In: Proceedings of the 4th international defense and homeland security simulation workshop (DHSS 2014), pp 13–20
12. Jedari B, Xia F, Ning Z (2018) A survey on human-centric communications in non-cooperative wireless relay networks. IEEE Commun Surveys Tutorials 20(2)
13. Jeong W, Nof SY (2008) Performance evaluation of wireless sensor network protocols for industrial applications. J Intell Manuf 19(3):335–345
14. Khan FA, Imran M, Abbas H, Durad MH (2017) A detection and prevention system against collaborative attacks in mobile ad hoc networks. Future Gener Comp Sy 68:416–427
15. Liu Y, Nof SY (2008) Fault-tolerant sensor integration for micro flow-sensor arrays and networks. Comput Ind Eng 54(3):634–647
16. Liu Z, Suzuki T (2018) Using agent simulations to evaluate the effect of a regional BCP on disaster response. J Disaster Res 13(2):387–395
17. Lu L, Wang X, Ouyang Y, Roningen J, Myers N, Calfas G (2018) Vulnerability of interdependent urban infrastructure networks: equilibrium after failure propagation and cascading impacts. Comput Aided Civil Infrastructure Eng 33(4):300–315
18. Mattsson L-G, Jenelius E (2015) Vulnerability and resilience of transport systems—a discussion of recent research. Transp Res Part A Policy Practice 81:16–34
19. Monostori L, Valckenaers P, Dolgui A, Panetto H, Brdys M, Csáji BC (2015) Cooperative control in production and logistics. Ann Rev Control 39:12–29
20. Moran T, Molnar JJ, Desourdis RI, Kurgan WM, Cloutier J-F (2015) Speeding power restoration. In: IEEE international symposium on technologies for homeland security, HST
21. Nof SY (2007) Collaborative control theory for e-work, e-production, and e-service. Ann Rev Control 21(2):281–292
22. Nof SY, Ceroni J, Jeong W, Moghaddam M (2015) Design with collaborative control theory. In: Revolutionizing collaboration through e-work, e-business, and e-service. Springer, Berlin, pp 33–75

23. Pérez ATE, Camargo M, Rincón PCN, Marchant MA (2017) Key challenges and requirements for sustainable and industrialized biorefinery supply chain design and management: a bibliographic analysis. Renew Sust Energ Rev 69:350–359
24. Radakov DV (1973) Schooling in the ecology of fish. Wiley and Israel Program for Scientific Translations, Jerusalem
25. Rajan VN, Nof SY (1996) Minimal precedence constraints for integrated assembly and execution planning. IEEE Trans Robot Autom 12(2):175–186
26. Rajan VN, Nof SY (1996) Cooperation requirement planning for multiprocessors: optimal assignment and execution planning. J Intell Rob Syst 15:419–435
27. Rege A (2016) Incorporating the human element in anticipatory and dynamic cyber defense. In: IEEE international conference on cybercrime and computer forensic (ICCCF 2016), 9 Nov 2016
28. Reimann F, Kosmol T, Kaufmann L (2017) Responses to supplier-induced disruptions: a fuzzy-set analysis. J Supply Chain Manage 53(4):37–66
29. Reyes Levalle R, Nof SY (2015) Resilience by teaming in supply network formation and re-configuration. Int J Prod Econ 160:80–93
30. Seok H, Kim K, Nof SY (2016) Intelligent contingent multi-sourcing model for resilient supply networks. Expert Syst Appl 51:107–119
31. Stewart GL, Barrick MR (2000) Team structure and performance: assessing the mediating role of intrateam process and the moderating role of task type. Acad Manage J 43(2):135–148
32. Stone P, Veloso M (1999) Task decomposition, dynamic role assignment, and low-bandwidth communication for real-time strategic teamwork. Artif Intell 110(2):241–273
33. Van de Walle B, Comes T (2014) Risk accelerators in disasters: insights from the Typhoon Haiyan response on humanitarian information management and decision support. In: Proceedings of advanced information systems engineering, 26th International Conference (CAiSE 2014), pp 12–23
34. Velasquez J, Yoon S, Partridge B, Nof SY (2005). Organizational knowledge learning and decision support for emergency and security challenges. In: 18th international conference on production research, Salerno, Italy, Aug 2005
35. Velasquez JD, Yoon SW, Nof SY (2010) Computer-based collaborative training for transportation security and emergency response. Comput Ind 61(4):380–389
36. Yang TJ, Fan W (2016) Information management strategies and supply chain performance under demand disruptions. Int J Prod Res 54(1):8–27
37. Zhang S, Wong TN (2017) Flexible job-shop scheduling/rescheduling in dynamic environment: a hybrid MAS/ACO approach. Int J Prod Res 55(11):3173–3196
38. Zhong H, Nof SY (2015) The dynamic lines of collaboration model: collaborative disruption response in cyber–physical systems. Comput Ind Eng 87:370–382
39. Zhong H, Nof SY, Filip FG (2014) Dynamic lines of collaboration in CPS disruption response. In: Proceedings of the 19th IFAC World Congress, Cape Town, South Africa
40. Zhong H, Wachs JP, Nof SY (2014) Telerobot-enabled HUB-CI model for collaborative lifecycle management of design and prototyping. Comput Ind 65(4):550–562
41. Zhong H, Nof SY, Berman S (2015) Asynchronous cooperation requirement planning with reconfigurable end-effectors. Robot Comput Integr Manuf 34(8):95–104

Chapter 3
The DLOC Model

3.1 Network Theory Foundation

E-Work systems are cyber-physical systems (CPS) that have many cyber and physical dependencies implemented by their links. For instance, if one element of a power grid fails, dependent facilities will stop functioning due to lack of power, and if a given control center is disconnected from the control network, the related power stations will refuse to work. This behaviors of inter-connection and inter-dependency is often modeled through complex networks.

3.1.1 Complex Networks and Interdependent Models

Frequently used network models for CPS are ER: Erdős-Rényi random graph [17]; BA: Barabási-Albert scale-free networks [5]; and WS: Watts-Strogatz small-world model [41].

In ER, the network connection is completely random, as each node is connected to any other node with an equal probability. In BA, however, the probability of having a highly connected node decreases exponentially. BA puts the emphasis on capturing the evolution of networks in nature: New nodes are added into a system with higher likelihood to be linked with high-degree neighbors. WS models have small average shortest path, but high clustering coefficients. The two properties enable WS to mimic the behavior of real networks, which are often clustered in many small groups. A comprehensive comparison of the network models can be found in Albert and Barabási [1].

Realizing that cyberspace and the physical world are two interdependent networks, researchers start to use two (sometimes more) interdependent networks to model complex systems. Interdependent networks belong to a class of models called

© Springer Nature Switzerland AG 2020
H. Zhong and S. Y. Nof, *Dynamic Lines of Collaboration*,
Automation, Collaboration, & E-Services 6,
https://doi.org/10.1007/978-3-030-34463-4_3

multilayer networks, or network of networks, which is the next frontier in network science [24]. In Buldyrev et al. [7], network robustness of interdependent networks is modeled as the existence of the remaining giant component after random attacks. It successfully captures the cascading effect occurring in a CPS. Based on this property and model, design considerations through simulations are provided [44]; the authors recommend the strategy of adding bi-directional links regularly to every node in each network (deterministically allocate each node exactly the same number of inter-edges) to increase robustness against random failures.

In addition, Wang et al. [40] have provided analysis on the influence of load across interdependent networks to improve the robustness of the networks against cascading failures. Despite the significant contributions in modeling CPS as interdependent networks, the models above are not sufficiently able to express the disruption response activities in CPS, especially when external resources are part of the e-Work and necessary to handle disruptions. The following subsections address two major needs that are required in the modeling of N2N services.

3.1.2   The Response Operations During Cascading Failures

Critical infrastructures play a fundamental role in our society. Because of e-Work, Cyber-Physical Infrastructures (CPIs) are emerging for all critical sectors. Once disruptions propagate over cyber or physical connections, the damage may cause a major disaster to humanity. The propagation of disruptions is often modeled as cascading failures in complex network theory [2, 30]. For interdependent networks, current active research has focused on the spreading process [19] and the attacks from one network to another network [35] The study on defensive and protective operations from one network to others is lacking.

In the current work, the responsibility of service agents in N2N services can be viewed as the response operations to eliminate the immediate and short-term effects of CPI disruptions. These response operations include diagnosis of the disruption, stabilization of the disruption, repairing the disruption, and preventing the disruption from propagation. The goal is to reduce damage to properties, and to minimize system down time.

The recent decade has seen a major increase in literature exploring the emergency management and disaster relief models and systems from different angles [34]. Most of the significant related works, e.g., Yi and Özdamar [45], focus on post-disaster relief problems, including evaluation operations, resource dispatching, etc. The response operations, however, during ongoing disruptions, are not well studied [14]. Several strategies on how to respond to cascading failures have been presented by Buzna et al. [8], considering network structure, response time delay, and the overall disposition of resources. The network representation by Buzna et al. [8], however, has a severe incompatibility if used to model disruption response in CPIs: The response time (from the start of disruption to the end of repair) is assumed to be independent of the availability of external resources. In addition, the resources deployed cannot be reassigned to other disruptions. These assumptions are not practical in the case of CPI disruption response. Consider the case when several power stations are failed in a cascading manner. A team of service agents, as external resources, needs to be deployed to the failed stations. If the resources are limited, the agents have to resolve the problems one by one, and thus the response time is dependent on the availability of resources. Besides the response time, the agents can be repeatedly assigned to new tasks.

In previous research, e.g., Chen and Nof [12], the constraints for errors and conflicts in large systems are modeled as complex networks. This work provides centralized and distributed algorithms to monitor and re-form the constraint network to detect, prioritize, and prevent propagating errors. The fitting of evolutionary constraint networks to conventional network models, however, requires further study. In addition, detection agents may be assumed to be deployed at each node, which relaxes the resource availability constraint and thus the dynamic collaboration behavior needs further investigation.

3.1.3 Link Ruptures and Concurrent Collaboration

Most of previous research in network science focuses on node failures. Similar to node failures, link ruptures can also cause cascading catastrophes. Research on link attacks (e.g., [31]) aims to identify vulnerabilities in existing networks and design new networks. Optimal link restoration procedures after disruptions in interdependent infrastructures are studied to identify the critical links that need to be restored and the sequence of restoration tasks [32, 36]. These link-based models seize the important role of links in networks but have provided limited insights on who we should respond to the disruptions and what strategies can improve their operational efficiency.

Link rupture is a useful tool in modeling collaborative disruption response (CDR). The uniqueness is that the link rupture can define a disruption that requires a concurrent collaboration of two agents providing services at two different but related sites. CDR is common in CPI, and it remains at the top of the current research agenda [23]. For example, in a freeway system, an incident on the road requires police officers to handle the situation. At the same time, several collaborating officers work at several different locations to close lanes or entrances, or to reverse flows. The collaboration is only achievable when agents (i.e., police officers in this case) are trained and have the collaboration technology to establish lines of command and collaboration during emergent situations [39].

In more advanced transportation systems, these operations will be handled by automated equipment, but the control logic is the same. Another class of needs for concurrent collaboration is due to the complexity in response activities, which often include diagnosing, repairing, restoring, and testing work of multiple elements simultaneously. The distributed collaboration is implemented through various channels from basic telephone communication, video-conferencing, to advanced concurrent coding and debugging [18].

Smart collaboration can optimize workflows and increase resource efficiency. Collaborative design and collaborative telerobotics are both cases that need experts to solve hard problems collaboratively [47]. Modeling of concurrent collaboration through link ruptures and repairing is one of the major focuses of this book.

3.2   Scheduling Theory

Determining which N2N service tasks should be performed by which service agents is a class of scheduling problems that aim to assign interdependent tasks to networked resources.

3.2.1   Traveling Repair-Agent Problem

In the research of emergency logistics management, the general problem of agents to repair a set of broken sites is often modeled as the Traveling Repair-agent Problem (TRP[1]), or the minimum latency problem [3, 9, 28]. For a given set of nodes, a repair-agent starts from a depot, visits all the nodes, and then comes back to the depot. The traveling of multiple agents among the nodes can be viewed as the process of reconfiguration.

TRP is a variation of the traveling salesman problem with a customer-centric objective function: Minimizing the total response time (or latency) for the nodes to be repaired. TRP is known to be NP-hard for arbitrary graphs. Many researchers have been investigating TRP and its variations. Recent reviews of taxonomy and solution approaches can be found in [16, 25, 29].

TRP attributes related to the DLOC work include:

1. *Multi-depot*: repair-agents are initially located at several service centers.
2. *Multi-agents*: multiple repair-agents are available for disruption response.
3. *Online scheduling*: service demands arrive while the agents are in service.
4. *Dynamic services*: services requested may change during the response operations.
5. *Operation synchronization*: two or more agents must execute their respective operation at their respective locations at the same time.

One interesting application of TRP is in the deploying and dispatching for emergency medical service (EMS). Current research has focused on how to route ambulances to respond to isolated calls from patients [20, 26]. Though several solution approaches have been proposed for different TRP variations, none of the models capture all the necessary characteristics of N2N services: Cascading service requests and the concurrent repair collaboration.

---

[1] Traditionally, "R" in TRP represents "repairman". For N2N, the service team could be a group of autonomous humans/robots/systems, and some of them conduct operations in cyber space. Repair-agent is more appropriate for the current work.

Besides determining the routes of the tours, there are several other questions that need to be answered in the models related to TRP. One of them is where the depots for the repair-agents should be located. The general facility location problem has been studied in various settings, e.g., Jia et al. [22] optimize the locations of medical suppliers. A thorough and general review can be found in Snyder [38]. Again, current research in literature is insufficient for handling N2N services, where the services are propagating, and concurrent collaboration is necessary.

3.2.2 Integrated Network Restoration and Scheduling Problem

How service agents make decisions and perform in the period during and immediately after a disruption is crucial to mitigating the influence of the disruptions. In literature, scheduling problems in post-disruption operations are under current research. Ang [4] has modeled the post-disruption restoration of an electrical power grid with a mixed-integer program. The repair schedule minimizes the cost of power-shedding during the restoration operations. Xu et al. [43] have applied genetic algorithm to schedule the inspection, damage assessment, and repair tasks to restore a power system after earthquake. Nurre et al. [32] and Cavdaroglu et al. [10] have established an integrated network restoration and scheduling model that focuses on identifying the critical arcs that need to be restored and the schedule of restoration tasks assigned to response agents.

Interdependencies between the client network components impose constraints on the repair schedule that no dependent node can be repaired without restoring parent nodes. The models mentioned above can be classified as parallel machine scheduling problems with job sequence limitations. This class of problems is NP-hard and often impossible to obtain the optimal solution efficiently.

Several heuristic solutions have been developed based on the shortest processing time (SPT) dispatch rule [33]. The method relies on the assumptions that

1. service agents work in parallel, and are independent of each other; and
2. the reconfiguration time between services are ignored.

Although the SPT rule is applicable to N2N services, the performance of scheduling is pending to be improved. Maya Duque et al. [27] have solved the problem of scheduling and routing for road (i.e., links in a graph) repair crew optimizing the accessibility after disasters. The authors have proposed an exact dynamic programming algorithm and an iterative greedy-randomized heuristic solution, but the solutions only work for networks with relatively small size (less than 50 nodes for the exact algorithm, and less than 500 nodes for the approximation heuristics). New approaches are required to handle N2N services with thousands or millions of nodes, which are typically present in e-Work systems.

3.3   The DLOC Framework and Model for Disruption Handling and Control

3.3.1   Model Assumptions

To define a reasonable scope for the current research, the following assumptions are made in developing the DLOC model and control protocols.

1. Clients: Clients are people, systems, or other agents. They have two states: 0 for not requesting service, and 1 for requesting a service. For example, an electrical component in a power grid is a client which requests a service when it is failed and requires a repair. Clients are linked into a network through their dependencies or interactions.
2. Propagating services: Service requests from clients are inter-related. Service on one client may trigger services on the other clients. The propagation is determined by the network structure of clients.
3. Servers: Services are fulfilled by service providers only for requests out of the client network. The requests from clients are instantly received by the servers, and the servers have to perform reconfiguration before providing services if their configuration is not aligned with the service requests. For instance, repair-agents have to travel (i.e., reconfigure) from depots to the service-requesting clients. It is assumed that services require a certain amount of time to be accomplished, and the services provided are error-free.
4. Server collaboration: Certain services require concurrent collaboration by servers. For instance, a leak on a pipeline, which connects two pump stations, requires two repair-agents working on the two stations to locate and to resolve the problem. To harvest fruits with reconfigurable end-effectors require the collaboration of grippers and finger tools. Not all servers can collaborate with each other. The servers are only linked to their eligible partners to form the server network.
5. Control protocols: A control protocol determines how the servers collaborate to serve the clients. Example protocols include how to locate the service depots, how to build the service network, and how to respond to service requests. Clients, however, are assumed to be uncontrollable directly, though there are influenced by servers' operations.

Based on these five assumptions, the detailed mathematical definitions of server and client networks, and the DLOC model are illustrated in the following section.

3.3.2   Model Components

The DLOC model has four major building blocks: the client network, the service propagation, the service team, and the failure prevention by agents.

A client network is defined as $G = (N, L)$ with link distribution P_G, where N is a set of components connected through a set of links (L). A link (i, j) represents dependency between the two linked nodes $(i, j \in N)$. To reduce the complexity at this level of abstraction, the links are modeled as undirected and un-weighted.

Each node or link in G has two states at any time step: 0 or 1. In '0', the element is active—not requesting service, while '1' represents disrupted—requesting service.

Disruptions that occur at time t are categorized into two types: link rupture and node failure.

$$\text{Client network link rupture:} \quad \text{state}_t(i, j) = 1 \text{ and state}_{t-}(i, j) = 0 \qquad (3.1)$$

$$\text{Client network node failure:} \quad \text{state}_t(i) = 1 \text{ and state}_{t-}(i) = 0 \qquad (3.2)$$

Let F be the set of disruptions that occur to the client network. The mapping from node failures and link ruptures to F is as follows: a node failure is defined as a single disruption $(i \in F)$, and a link rupture is represented as two coupled disruptions $(i, j \in F)$.

Each disruption is uniquely defined by a 3-tuple $(i = \langle \tau_i, \upsilon_i, \gamma_i \rangle, i \in F)$, where τ_i is the initial timestamp of this disruption $(\tau_i \geq 0)$; υ_i represents the location of the disruption $(\upsilon_i \in N)$. Note: here i is *not* the index of nodes in G but an element of F); γ_i is a rupture reference defined in Eq. (3.3).

$$\text{Link rupture reference:} \quad \gamma_i = \begin{cases} j, & \text{if state}_{\tau_i}(\upsilon_i, \upsilon_j) = 1 \text{ and state}_{\tau_i^-}(\upsilon_i, \upsilon_j) = 0 \\ \emptyset, & \text{if state}_{\tau_i}(\upsilon_i) = 1 \text{ and state}_{\tau_i^-}(\upsilon_i) = 0 \end{cases}$$

$$(3.3)$$

The above equation shows that if two disruptions $(i$ and $j)$ are used to represent a link rupture together, the rupture references of them points to each other. If a disruption is used to represent a node failure, the reference is an empty pointer.

Repair-agents (i.e., disruption responders) are initially located at depots (D). Each agent has its own initial depot. Following the Hakimi property [21], the locations of a depot should be on the nodes of the client network $(i \in D \Longrightarrow i \in N)$. With this definition, a node in the client network can be represented as the depot for more than one agent (e.g., $i, j \in D, i = j = $ 'node k' $\in N$). In other words, even though some agents are deployed at the same physical location, distinguishable notations are used to represent the depots. This formulation has advantages in:

1. All depots have a uniform capacity 1; and
2. only binary decision variables are required.

For this model, D and F establish a virtual graph $(D \cup F, \{c_{ij}, i, j \in D \cup F\})$ with directed and weighted links $(\{c_{ij}\})$. The repair-agents travel on $\{c_{ij}\}$ to handle disruptions. Each link weight of the virtual graph has two parts: the traveling time between two locations, and the repair timespan used to remove disruptions, as shown in Eq. (3.4).

Weights of virtual graph: $c_{ij} = t_D(i, j) + t_R(j) = d(i, j)/v + t_R(j)$ (3.4)

where $t_D(i, j)$ and $d(i, j)$ are the traveling time and a distance function between the represented nodes in the client network. For example, if $i \in D$ and $i =$ 'node k_1' $\in N$, the represented node for i is 'node k_1'. If $j \in F$ and $v_j =$ 'node k_2' $\in N$, the represented nodes for j is 'node k_2'. Thus, $d(i, j)$ is the distance between k_1 and k_2.

The distance function is application specific. It can be Euclidean distance, shortest paths, or any distance measures in any space suitable for the application. v is the velocity of an agent traveling in the same space with d. $t_R(j)$ is the time required to remove the disruption j, if $j \in F$. Otherwise, if $j \in D$, $t_R(j) = 0$. If j is part of a link rupture ($\gamma_j = k$), $t_R(j) = t_R(k)$, which shows the current collaboration lasts the same time for the repair at two different sites.

Then, the collaborative TRP with cascading failures is formulated as Eqs. (3.5)–(3.14).

$$\text{DLOC objective function:}\quad \min z = \sum_{i \in F} (\sigma_i - \tau_i) \qquad (3.5)$$

$$\text{s.t.}\quad \text{Response sequence constraint:}\quad \sigma_j \geq \sum_{i \in D \cup F} x_{ij} c_{ij} + \sum_{i \in F} x_{ij} \sigma_i, \quad \text{for } j \in F$$
$$(3.6)$$

$$\text{Depot visit constraint:}\quad \sum_{j \in F} x_{ij} = \sum_{j \in F} x_{ji} \leq 1 \quad \text{for } i \in D \qquad (3.7)$$

$$\text{Disruption visit constraint:}\quad \sum_{j \in D \cup F} x_{ij} = \sum_{j \in D \cup F} x_{ji} = 1 \quad \text{or } i \in F \qquad (3.8)$$

$$\text{Depot-wander elimination constraint:}\quad \sum_{i,j \in D} x_{ij} = 0 \qquad (3.9)$$

$$\text{Sub-tour elimination constraint:}\quad \sum_{i,j \in S} x_{ij} \leq |S| - 1 \quad \text{for } S \subseteq F \qquad (3.10)$$

$$\text{Service constraint:}\quad \sigma_i \geq \tau_i \quad \text{for } i \in F \qquad (3.11)$$

$$\text{Concurrent collaboration requirement:}\quad \sigma_i = \sigma_j \quad \text{for } \gamma_i = j, \gamma_j = i, \text{ and } i, j \in F$$
$$(3.12)$$

$$\text{Decision variables for service tours:}\quad X = \{x_{ij} = \{0, 1\}, i, j \in D \cup F\} \quad (3.13)$$

$$\text{Cascading function of disruptions:}\quad F = CASCAD(F_0, G, X) \qquad (3.14)$$

where

z is the target value of the objective function;
σ_i is the time of finishing the repair of disruption i;

τ_i is the start time of disruption i; x_{ij} is a binary decision variable indicating if a repair-agent should go from node i to node j (i, j are either a depot or a disruption) and X is the set of all x_{ij};

S is any subset of disruptions;

F_0 is the set of disruptions at time 0; and

$CASCAD$ is a model that generates the entire disruption set (F) based on initial disruptions.

Equation (3.5) is the objective function to minimize the response time of all disruptions. From the perspective of scheduling theory, the objective function is to minimize the total completion time.

Another possible objective function is minimizing the makespan, which is the time to fulfill all services requests. The makespan is optimized in the related problem: traveling salesman problem.

Due date related objective functions, such as tardiness, are not considered because the due date is not usually specified in this type of problems.

In the case of response to disruptions, the makespan is the time that the entire system is recovered (t_{recover}).

Though the recovery time is an important performance indicator to indicate the finishing point of the N2N services, it is relatively more critical to fulfill each customer's request as soon as possible. The current research, therefore, uses the customer-centric total completion time as the objective function.

The response sequence constraint (in Eq. (3.6)) ensures that, in a repair tour, the restore time of a disruption should be larger than the restore time of the previous node in the same tour, plus the traveling time from the previous node to the current node. The proof is as follows.

Proof of Response Sequence Constraint (in Eq. (3.6)) Suppose a disruption node $j \in F$ is handled by a repair-agent a. Let $k \in D \cup F$ be the node visited by a right before j. $x_{kj} = 1$ because c_{kj} is used in the routes for disruption response. And $x_{ij} = 0$ for $i \neq k, i \in D \cup F$.

$$\sum_{i \in D \cup F} x_{ij} c_{ij} = x_{kj} c_{kj} = c_{kj} \tag{3.15}$$

$$\sum_{i \in F} x_{ij} \sigma_i = x_{kj} \sigma_k = \sigma_k \tag{3.16}$$

The finishing time of repairing node j should be larger than or equal to the finishing time of node k plus the traveling time from k to j and the repair time span of node j. The inequality is included because agent a may have to wait for another agent to perform a concurrent collaboration for repairing a link rupture.

$$\sigma_j \geq \sigma_k + c_{kj} \tag{3.17}$$

Response sequence constraint (in Eq. (3.6)) is proven.

Depot visit constraint (in Eq. (3.7)) shows that if a repair-agent leaves a depot, it must come back to a depot.

Disruption visit constraint (in Eq. (3.8)) determines that each disruption node has to be visited by one and only one repair-agent.

Wandering, or a repair tour without visiting any disruption node, is not allowed (depot-wander elimination constraint, in Eq. (3.9)). Equation (3.10) is the sub-tour elimination constraint that a tour without visiting depots is not valid.

Service constraint (in Eq. (3.11)) ensures the repair of disruptions must occur after the disruption.

The collaborative repair on links is implemented according to the concurrent collaboration requirement (in Eq. (3.12)) as the repair time is the same for the two disruptions representing a single link rupture.

Equation (3.13) shows the scope of decision variables for repair tours.

Cascading function of disruptions (in Eq. (3.14)) shows the entire disruption set is generated according to certain cascading function, taking as input the initial failures, the structure of the client network, and the sequence of repair.

In the model for DLOC, the set of total disruptions is determined by a cascading process (the *CASCADE* function in Eq. (3.14)), i.e., the service requests propagate over the client network. Many models have been developed in the literature that try to imitate this process based on the different characteristics of systems. The models can be categorized into two classes:

- Structure-based models, and
- load-based models.

The first class is based on the assumption that failed elements will cause structurally connected components to fail. Percolation theory is often used to model the cascading process [6, 42].

The second class of models assumes that the original load on a failed element should be redistributed to other elements, which would cause additional failures due to overload [13, 30]. In the current work, an adapted Watts cascade model [42] is used to indicate the cascading process in the client network. Watts has proven that this structure-based model is suitable for modeling the cascading failures in large scale systems, such as power grids and Internet. The advantage of Watts model is that it only relies on the topological information, and does not require additional assumptions of loads in the target system. The Watts model is, therefore, appropriate in the current research for general e-Work systems.

The details of the *CASCADE* service propagation procedure are as follows. At the initial time, all links and nodes are in 0 states. When a node or a link is failed, its state is switched to 1. The cascading of failures is modeled as a sequence of state

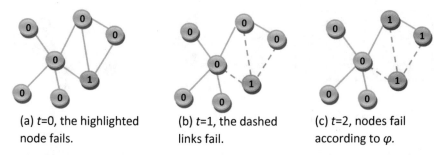

(a) $t=0$, the highlighted node fails.

(b) $t=1$, the dashed links fail.

(c) $t=2$, nodes fail according to φ.

Fig. 3.1 Example of cascading failures ($\varphi = 0.3$)

changes. If no remediation is present ($x_{ij} = 0, \forall\, i, j \in D \cup F$), after a node failed at time t, all connected links will then be deactivated at $t + 1/u$, where u is the spread speed of failures (measured by number of links per time unit). At each time step, an active node will determine its state by assessing its connected links.

If at least φ fraction of its links are in state 1, the node adopts state 1, otherwise it retains its state 0 ($0 < \varphi \le 1$). Links and nodes remain failed after being disrupted, unless repaired by service agents. A cascading failures example of three starting steps is shown in Fig. 3.1. Depending on the local network structure, an initial failure fades away quickly, or propagates to a system collapse. The example failure process shown in Fig. 3.1 should end at time $t = 6$ with no surviving element.

3.3.3 Service Team

External to the modeled client network (G), the service agents form another network defined as the service team: $R = (A, C)$ with total number of service agents n_A. A is a set of agents that can restore failed nodes and links in client network. C shows the collaboration-ability between every two agents in R. If two agents are able to collaborate (according to their responsibility, skills, and workflow), the two agents are connected in R; otherwise the two agents are not connected.

Each agent has two states: 0 for idle, and 1 for working. Initially, each agent has an inter-edge (E) with one node in G (i.e., the depot). If a node is failed, an agent will be assigned to repair, and thus transits to state 1. The working agent disconnects itself from its depot (i) or current location and connects with the failed node (j). It is assumed that the travel of an agent will not be affected by the failures in the client network. Therefore, agents can link to failed nodes even if some links or nodes between i and j are failed. This assumption is reasonable for client networks, because the response team should have backup means to complete the response task. For example, to restore failed power stations and power lines, service agents can travel on highways which are not part of the smart grid network.

The collaboration-ability (C) values are specifically designed for N2N services. A link (i, j) is only repairable when (1) two agents are connected with each vertex

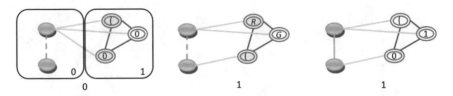

Fig. 3.2 Example of link repair

of the link, and (2) the two agents are able to collaborate in the response team (R).
This procedure is illustrated in Fig. 3.2.

After nodes and links are restored to normal state, the agents are back to state 0.
It is assumed that the agents and their collaboration abilities are free of disruptions.
While working or traveling, agents cannot be interrupted to serve on new tasks.
New tasks are appended to the current tasks. (These simplifying assumptions will be
relaxed in future work.)

3.3.4 Proactive Service by Agents

An important advantage thanks to the presence of service agents is that agents are
able to prevent errors from propagating in the modeled client network. If an agent
is supervising the element which is being affected by cascading failures, the agent
can prevent the error from occurring. Supervision, node failure prevention, and link
rupture prevention are defined as follows.

A node (i) is supervised if and only if there is an inter-edge between this node
and an agent.

Client network node supervision:

$$\text{supervision}(i) = \begin{cases} 1 & \exists (i, a) \in E \\ 0 & otherwise \end{cases}, \quad i \in N, a \in A \qquad (3.18)$$

A link (i, j) is supervised if and only if two vertexes of the link are supervised and
the supervising agents are able to collaborate.
Client network link supervision:

Client network link supervision:

$$\text{supervision}(i, j) = \begin{cases} 1 & \exists (i, a) \in E, (j, b) \in E, (a, b) \in C \\ 0 & otherwise \end{cases}, \quad i, j \in N, a, b \in A$$
$$(3.19)$$

As modeled by the cascading failures, a node will fail if at least φ fraction of its
edges are failed. If one failed link is supervised, however, this link does not count

as a failed link to advance the node failure. The definition of node failure is updated as follows: A node will only fail, if at least φ fraction of its edges are failed and unsupervised (in Eq. (3.20)).

A link rupture with supervision is relatively more complicated. If only one of its vertexes is failed, the link will fail as the failed node is not supervised. If both vertexes are failed, this link will fail unless supervised. Link rupture is redefined in Eq. (3.21).

Node failure in client network (with failure prevention):

$$state(i) = \begin{cases} 1 & \frac{\sum_{(i,j)\in L} state(i,j)-\sum_{state(i,j)=1} supervision(i,j)}{deg_G(i)} \geq \varphi \\ 0 & otherwise \end{cases}, \quad i, j \in N \quad (3.20)$$

Client network link rupture (with failure prevention):

$$state(i, j)$$
$$= \begin{cases} state(i) \times (1 - supervision(i)) + state(j) \times (1 - supervision(j)) & state(i) \neq state(j) \\ supervision(i, j) & state(i) = state(j) = 1, \quad i, j \in N \\ 0 & state(i) = state(j) = 0 \end{cases}$$
$$(3.21)$$

It is clear that if all nodes and links in the client network are supervised, the client network will be free of errors. But in reality, the response resources may not be adequate. Based on the mathematical model for DLOC, N2N service managers are challenged by three fundamental questions:

1. How should the service team be assembled (i.e., planning the team structure and team size)?
2. Where should the depots be?
3. What response workflow protocol needs to be used?

The following chapters provide solutions to these three questions.

3.4 Model Comparison

Existing models found in literature cannot completely capture the combination of properties in network-to-network (N2N) service problems. Table 3.1 lists all required elements for constructing the new model for DLOC and contributing models previously developed in literature. DLOC combines the network science, scheduling theory in operations research, and collaborative control theory (CCT) to provide an analysis and solution for the N2N problem. Table 3.2 summarizes the performance metrics commonly used in related models. There are three classes of metrics: (1) time related, (2) cost related, and (3) quality related. The metrics will guide the design of experiments for this work, as described later in this book.

Table 3.1 Summary of related models for DLOC

Related models		Properties required						
Model category	Model (example references)	Service team	Nodes failure	Link rupture	Cascading disruptions	Online service	Response protocol	Concurrent collaboration
Network science	Complex network with node attacks [30]	–	Yes	–	Yes	–	–	–
	Complex network with links attacks [31]	–	Yes	Yes	Yes	–	–	–
	Interdependent networks [7, 44]	partial	Yes	–	Yes	–	–	–
	Response strategies to cascading failures [8]	Partial	Yes	–	Yes	Yes	Partial	–
Scheduling theory	TRP [16, 29]	–	Partial	–	Partial	Partial	Yes	Partial
	Facility location problem [22, 38]	Partial	–	–	–	Partial	–	–

(continued)

Table 3.1 (continued)

Related models		Properties required						
Model category	Model (example references)	Service team	Nodes failure	Link rupture	Cascading disruptions	Online service	Response protocol	Concurrent collaboration
	Integrated network design and scheduling [33] (Cavdaroglu et al. [10])	Partial	Yes	–	–	Yes	Partial	–
CCT	Collaborative intelligence for collaborative decision making [15, 46]	Yes	Yes	–	–	–	Partial	Yes
	Conflict-error detection and prevention [12, 11]	–	Yes	–	Yes	Yes	Partial	–
	Preliminary DLOC [48]	Partial	Yes	Yes	Partial	Yes	Partial	Yes
	DLOC (this work)	Yes	Yes	Yes	Yes	Yes	Yes	Yes

"–": not yet available

Table 3.2 Metrics Composition

Metric class	Descriptions	Example references
Time	Total response time, the maximum latency	[8, 29]
Cost	Minimum required agents, distance travelled by agents, equipment cost, total damage, failed nodes, utilization	[7, 37, 48]
Quality	Conflict/error preventability, quality-of-coverage of demand area, preparedness, recovery rate, productivity	[12, 11, 22, 26]

References

1. Albert R, Barabási AL (2002) Statistical mechanics of complex networks. Rev Mod Phys 74(1):47–97
2. Albert R, Jeong H, Barabási A (2000) Error and attack tolerance of complex networks. Nature 406:378–382
3. Altay N, Green WG III (2006) OR/MS research in disaster operations management. Eur J Oper Res 175(1):475–493
4. Ang CC (2006) Optimized recovery of damaged electrical power grids (Doctoral dissertation, Monterey, California. Naval Postgraduate School)
5. Barabási A-L, Albert R (1999) Emergence of scaling in random networks. Science 286(5439):509–512
6. Bashan A, Parshani R, Havlin S (2011) Percolation in networks composed of connectivity and dependency links. Phys Rev E 83(5):051127
7. Buldyrev SV, Parshani R, Paul G, Stanley HE, Havlin S (2010) Catastrophic cascade of failures in interdependent networks. Nature 464:1025–1028
8. Buzna L, Peters K, Ammoser H, Kuhnert C, Helbing D (2007) Efficient response in cascading disaster spreading. Phys Rev E 75:056107
9. Caunhye AM, Nie X, Pokharel S (2012) Optimization models in emergency logistics: a literature review. Socio-Econ Plan Sci 46(1):4–13
10. Cavdaroglu B, Hammel E, Mitchell JE, Sharkey TC, Wallace WA (2013) Integrating restoration and scheduling decisions for disrupted interdependent infrastructure systems. Ann Oper Res 203(1):279–294
11. Chen XW, Nof SY (2012) Conflict and error prevention and detection in complex networks. Automatica 48(5):770–778
12. Chen XW, Nof SY (2012) Agent-based error prevention algorithms. Expert Syst Appl 39(1):280–287
13. Crucitti P, Latora V, Marchiori M (2004) Model for cascading failures in complex networks. Phys Rev E 69(4):045104
14. Day JM (2014) Fostering emergent resilience: the complex adaptive supply network of disaster relief. Int J Prod Res 52(7):1970–1988
15. Devadasan P, Zhong H, Nof SY (2013) Collaborative intelligence in knowledge based service planning. Expert Syst Appl 40(17):6778–6787
16. Drexl M (2012) Synchronization in vehicle routing—a survey of VRPs with multiple synchronization constraints. Transport Sci 46(3):297–316
17. Erdős P, Rényi A (1959) On random graphs, I. Publicationes Mathematicae (Debrecen) 6:290–297
18. Estler HC, Nordio M, Furia CA, Meyer B (2013) Collaborative debugging. In: Proceedings of 8th international conference on global software engineering, pp 110–119

19. Gao J, Buldyrev SV, Stanley HE, Havlin S (2012) Networks formed from interdependent networks. Nat Phys 8(1):40–48
20. Gnanasekaran AM, Moshref-Javadi M, Zhong H, Moghaddam M, Lee S (2013) Impact of patients priority and resource availability in ambulance dispatching. In: Proceedings of 2013 industrial & systems engineering research conference (ISERC)
21. Hakimi SL (1965) Optimum distribution of switching centers in a communication network and some related graph theoretic problems. Oper Res 13(3):462–475
22. Jia H, Ordóñez F, Dessouky MM (2007) Solution approaches for facility location of medical supplies for large-scale emergencies. Comput Ind Eng 52(2):257–276
23. Jiang Y, Yuan Y, Huang K, Zhao L (2012) Logistics for large-scale disaster response: achievements and challenges. In: Proceedings of IEEE 45th Hawaii International conference on system science (HICSS), pp 1277–1285
24. Kenett DY, Perc M, Boccaletti S (2015) Networks of networks—an introduction. Chaos Solitons Fractals 80:1–6
25. Krumke SO, de Paepe WE, Poensgen D, Stougie L (2003) News from the online traveling repairman. Theoret Comput Sci 295(1–3):279–294
26. Lee S (2011) The role of preparedness in ambulance dispatching. J Oper Res Soc 62(10):1888–1897
27. Maya Duque PA, Dolinskaya IS, Sörensen K (2016) Network repair crew scheduling and routing for emergency relief distribution problem. Eur J Oper Res 248(1):272–285
28. Moshref-Javadi M, Lee S (2016) The latency location-routing problem. Eur J Oper Res 255(2):604–619
29. Moshref-Javadi M, Lee S (2013) A taxonomy to the class of minimum latency problems. In: Proceedings of 2013 industrial & systems engineering research conference (ISERC)
30. Motter AE, Lai Y (2002) Cascade-based attacks on complex networks. Phys Rev E 66:065102(R)
31. Motter AE, Nishikawa T, Lai Y (2002) Range-based attack on links in scale-free networks: are long-range links responsible for the small-world phenomenon? Phys Rev E 66:065103(R)
32. Nurre SG, Cavdaroglu B, Mitchell JE, Sharkey TC, Wallace WA (2012) Restoring infrastructure systems: an integrated network design and scheduling problem. Eur J Oper Res 223(3):794–806
33. Nurre SG, Sharkey TC (2014) Integrated network design and scheduling problems with parallel identical machines: complexity results and dispatching rules. Networks 63(4):306–326
34. Ortuño MT, Cristóbal P, Ferrer JM, Martín-Campo FJ, Muñoz S, Tirado G, Vitoriano B (2013) Decision aid models and systems for humanitarian logistics. a survey. In: Vitoriano B et al (eds) Decision aid models for disaster management and emergencies. Atlantis Press, Paris, pp 17–44
35. Podobnik B, Horvatic D, Lipic T, Perc M, Buldú JM, Stanley HE (2015) The cost of attack in competing networks. J R Soc Interface 12(112). https://doi.org/10.1098/rsif.2015.0770
36. Shen S (2013) Optimizing designs and operations of a single network or multiple interdependent infrastructures under stochastic arc disruption. Comput Oper Res 40(11):2677–2688
37. Shuang Q, Zhang M, Yuan Y (2014) Node vulnerability of water distribution networks under cascading failures. Reliab Eng Syst Safety 124:132–141
38. Snyder LV (2006) Facility location under uncertainty: a review. IIE Trans 38(7):547–564
39. Velasquez JD, Yoon SW, Nof SY (2010) Computer-based collaborative training for transportation security and emergency response. Comput Ind 61(4):380–389
40. Wang J, Wu Y, Li Y (2015) Attack robustness of cascading load model in interdependent networks. Int J Mod Phys C 26(3), 1550030-1-14
41. Watts DJ, Strogatz SH (1998) Collective dynamics of small-world networks. Nature 393(6684):440–442
42. Watts D (2002) A simple model of global cascades on random networks. Proc Nat Acad Sci USA 99(9):5766–5771

43. Xu N, Guikema SD, Davidson RA, Nozick LK, Çağnan Z, Vaziri K (2007) Optimizing scheduling of post-earthquake electric power restoration tasks. Earthq Eng Struct Dyn 36(2):265–284
44. Yagan O, Qian D, Zhang J, Cochran D (2012) Optimal allocation of interconnecting links in cyber-physical systems: interdependence, cascading failures and robustness. IEEE Trans Parallel Distrib Syst 23(9):1708–1721
45. Yi W, Özdamar L (2007) A dynamic logistics coordination model for evacuation and support in disaster response activities. Eur J Oper Res 179:1177–1193
46. Zhong H, Reyes Levalle R, Moghaddam M, Nof SY (2015) Collaborative intelligence—definition and measured impacts on internetworked e-Work. Manage Prod Eng Rev 6(1):67–78
47. Zhong H, Wachs JP, Nof SY (2014) Telerobot-enabled HUB-CI model for collaborative lifecycle management of design and prototyping. Comput Ind 65(4):550–562
48. Zhong H, Nof SY, Filip FG (2014) Dynamic lines of collaboration in CPS disruption response. In: Proceedings of the 19th IFAC World Congress, Cape Town, South Africa

Chapter 4
Protocols for the Dynamic Lines of Collaboration

4.1 Asynchronous Collaboration Requirement Planning: The Configuration Design of the Service Team Prepared to Handle Collaborative Tasks

The first research question about the DLOC model is to determine how to plan the service team. This question is illustrated with an example of designing automated harvesting system with reconfigurable end-effectors (REEs). Automated harvesting is a sophisticated process, because the environment is usually unstructured and the grasp quality is highly demanding [1].

The targets (e.g., fruits) form an interdependent client network, as the harvesting operations for one target (e.g., pruning) can affect the operations of other targets. Each part of REEs can be viewed as an agent, and the harvesting operations cannot be fulfilled without the collaboration between the agents. It is commonly accepted that more sophisticated tasks require greater flexibility. The reconfiguration of the REE underpins dynamic collaborations of part agents to provide network-to-network (N2N) services to the client network.

The Collaboration Requirement Planning (CRP) principle in Collaborative Control Theory (CCT) has been established to first design a static collaboration team (CRP-I), and then apply and re-plan it dynamically in real-time (CRP-II) for flexible manufacturing and robotic systems [14]. Two considerations motivated the adaptation of the CRP methodology to Asynchronous Collaboration Requirement Planning (ACRP):

(1) Stochasticity. Decisions about when to change and what kind of end-effector configuration to use, should be prepared for unknown yet predictable targets. Therefore, ACRP module should make dynamic decisions based not only on the configurations, but also on stochastic tasks.

© Springer Nature Switzerland AG 2020
H. Zhong and S. Y. Nof, *Dynamic Lines of Collaboration*,
Automation, Collaboration, & E-Services 6,
https://doi.org/10.1007/978-3-030-34463-4_4

(2) Asynchrony. Several arms can carry end-effectors at the same time to increase productivity. Those arms should work in parallel and asynchronously to adapt to a complex work environment.

Automated systems, harvesting platforms for illustration, have several parallel components. New requests of harvesting jobs arrive at the workstation independently of the system's operations. Cooperating manipulators may have different speeds and tasks. Besides, system modeling may be running at a virtual speed different from the actual manipulations (often faster). CRP planner agent should not wait till the end of the previous task to generate the plan for only the next job.

In the cases when job requests arrive faster than their completion, the planner agent must be agile to adjust plans suitable for both the current job and future jobs. In harvesting, for instance, a job is defined as the set of operations to pick and convey a fruit from a plant to a container. The job may also include an action to reconfigure end-effectors in order to have better grasp performance.

To address the challenges stated above, ACRP is developed (as illustrated in Fig. 3.3). The ACRP framework works in a centralized asynchronous multi-agent system (i.e., client/server architecture). The CRP planner (CRP-I and CRP-II) is the server agent which accepts requests from all other agents. The five parallel asynchronous agents are (Fig. 3.3):

1. Supervisor. The supervisor is a human planner/operator or a supervisory program/agent. The jobs assigned by the supervisor arrive at the automated platform in a stochastic sequence. The inter-arrival intervals and the categories of jobs are estimated by queuing models.
2. Cell control system. The automation platform is controlled under certain rules, unless the rules are changed by the CRP planner. For example, the velocity of a harvesting vehicle remains constant unless changes in the density of harvested fruits are detected.
3. Modeling systems. Potential costs for different tasks are estimated by the modeling systems which run with different scenarios independently. Once the CRP planner makes decisions, it uses the estimated cost information to produce simultaneous matching (assignments) and planning.
4. Robot arms. Each manipulator or robot carries a REE. They execute a sequence of tasks assigned by the CRP planner.
5. CRP planner. The CRP planner has two parts. The CRP-I part prepares a global plan for the entire system based on estimations. The plan determines the configurations (number of arms and interchangeable parts) needed to be mounted in the work station. The global plan also includes the policy for the cell movement. The CRP-II part determines which configuration to use online when actual requests from supervisors are received (Fig. 4.1).

The original CRP planner generates an optimal plan by searching all possible combinations of tasks for a given job. The CRP scheduling protocol is acceptable for stationary workstations with given jobs.

In comparison, the planning module in the new ACRP framework needs to make decisions for stochastic jobs detected by agents asynchronously. To handle

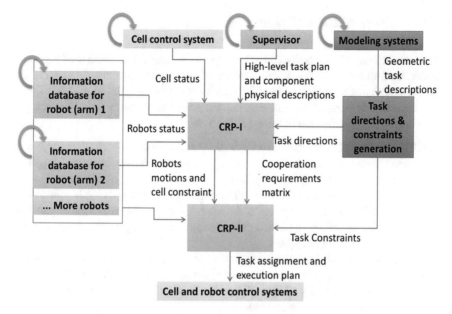

Fig. 4.1 Framework of asynchronous cooperation requirement planning (ACRP)

the stochasticity, ACRP improves both the design and the control of REE systems. In the design phase, ACRP selects the required configurations for the REE based on the estimation of relevant costs. In the online control phase, ACRP calculates an optimal, or nearly optimal solution to determine the configurations of handling both detected and predicted jobs. The solution is also updated when new jobs arrive stochastically.

The following sections address how to use ACRP to design REE for a harvesting process.

4.1.1 Reconfiguration Design

A REE system is determined by its configurations, each with its own function and ability of handling a set of objects. To design the reconfigurability of the automated harvesting system means to integrate necessary REE configurations into the system.

After the integration, the system should have a good production quality and yield, while maintaining a reasonable cost. As grasp quality is essential for automated harvesting, the design for reconfigurability focuses on the grasp cost. Integrated REE should have high grasp quality for a large range of items in the harvesting task. Besides grasp cost, reconfiguration cost is also minimized to improve efficiency. To present the reconfiguration design method, some preliminary definitions are required.

4.1.2 Order of Reconfigurability

The order of reconfigurability (OOR) is used to represent the number of reconfigurable layers in an automated, self-configured end-effector system. OOR indicates
the complexity of a REE. Higher OOR enables more reconfigurability, but it would
also cost more in terms of the time required for the reconfiguration task. For example,
a REE system is configured in three OOR levels:

- 1OOR: An end-effector module with the first order is defined as a REE that can only
  change the entire end-tool. For example, an automated harvesting arm changes its
  gripper to shears for pruning purposes.
- 2OOR: The second order includes all configurations in the first order, plus additional interchangeable components. An interchangeable finger is a good example.
  Different fingers can be mounted onto a gripper to have specific contact textures
  and forces for different objects.
- 3OOR: With the third order, a REE further changes the configuration details. For
  example, a three-finger gripper has two positions (gestures): The angles between
  the fingers are (1) equally 120° or (2) two 150° and one 60° to produce two different
  force directions on the grasped object.

In some rare and sophisticated cases, a REE has more configuration mechanisms,
so that a system can have an even higher OOR. 0OOR is defined for a dedicated
system without any reconfigurability.

4.1.3 Configuration Network of the Client Team

Increasing the OOR will increase the time and cost of reconfiguration during harvesting process. Figure 4.2 shows an example of the configuration network for a
harvesting system with 3OOR and three arms.

When a picking job is requested, the harvesting system will first determine which
arm to use. Then, the system further configures the required end-effector settings.
Reconfiguration design implies the determination of the structure of configuration
networks. For representation, if a configuration of an arm has gripper i, finger j, and
position m, the configuration is defined as $\langle i, j, m \rangle$.

4.1.4 Estimated Total Reconfiguration Cost

Estimated reconfiguration cost (C_{config}) represents the efforts required to execute the
reconfiguration operations, which include mounting and dismounting end-effector
parts from/to a tool rack, and the changes of end-effector positions. The reconfiguration cost should be considered because it is the major indicator of the adaptation

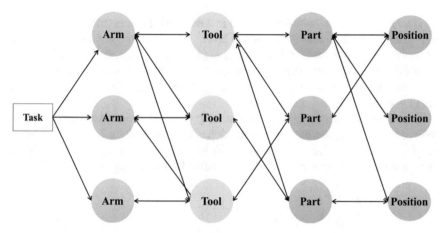

Fig. 4.2 Configuration network of a system with three arms and 3OOR

flexibility loss due to a large set of configurations. As the jobs are stochastic, the exact reconfiguration cost for a harvesting process cannot be obtained beforehand. Thus, estimation is necessary. Systems with relatively more configuration layers are more complex when they need to be reconfigured, so the cost for reconfiguration is relatively higher.

If a reconfiguration changes multiple layers, the reconfiguration cost is the sum of costs in all involved layers. The total amount of estimated reconfiguration cost (C_{config}) is a sum of costs of all transitions between any two possible configurations. C_{config} is calculated as follows.

$$\text{Total reconfiguration cost:} \quad C_{config} = \sum_{a,b \in M} f(a, b) c_{config}(a, b) \quad (4.1)$$

where

C_{config} is the total estimated reconfiguration cost;
M is the total number of configurations;
f is the frequency of switching between two configurations; and
c_{config} is the cost of switching between two configurations.

Generally, the reconfiguration cost is related to the layer of the configuration network due to the mechanical design of REE. Suppose $c_{config}(i, \cdot)$ is the reconfiguration cost in the ith layer of the configuration network, and M_i is the total number of parts in the ith layer ($M_i > 0$).

For example, $c_{config}(3, \langle 1, 2 \rangle)$ indicates the reconfiguration cost in the third layer for gripper No. 1 and finger No. 2 (proportional to the time used to change the gesture of finger 2 on gripper 1). $M_2(1)$ indicates the total number of reconfigurable fingers (i.e., layer 2) for gripper No. 1. If a transition only involves configurations in the third layer (from $\langle i, j, m \rangle$ to $\langle i, j, n \rangle$), only $c_{config}(3, \langle i, j \rangle)$ is calculated.

If a reconfiguration involves the third and the second layers (from $\langle i, p, m \rangle$ to $\langle i, q, n \rangle$), the total cost includes the second layer cost $c_{\text{config}}\,(2, i)$ and the average cost of the two corresponding third layer $((c_{\text{config}}\,(3, \langle i, p \rangle) + c_{\text{config}}\,(3, \langle i, q \rangle))/2)$.

If a transition involves all three layers (from $\langle k, p, m \rangle$ to $\langle l, q, n \rangle$), the cost in three layers should be summed together: $((c_{\text{config}}\,(3, \langle k, p \rangle) + c_{\text{config}}\,(3, \langle l, q \rangle))/2 + ((c_{\text{config}}\,(2, k) + c_{\text{config}}\,(2, l))/2 + c_{\text{config}}\,(1)$.

In practice, the frequency of reconfiguration (f) is not known in the design phase. Without this knowledge, the initial estimation of f can be a constant. Better estimation is achievable through iterative design and prototyping. By applying the layer-specific reconfiguration cost and constant reconfiguration frequency, Eq. (4.1) is rewritten as:

$$
\begin{aligned}
C_{\text{config}} = \sum_{i=1}^{M_1} \Bigg(\sum_{j=1}^{M_2(i)} & \Bigg[\binom{M_3(i, j)}{2} c_{\text{config}}(3, \langle i, j \rangle) \\
& + \sum_{q=p+1}^{M_2(i)} \frac{M_3(i, p) \cdot M_3(i, q)}{2} \left(\frac{c_{\text{config}}(3, \langle i, p \rangle) + c_{\text{config}}(3, \langle i, q \rangle)}{2} + c_{\text{config}}(2, i) \right) \Bigg] \\
& + \sum_{l=k+1}^{M_1} \Bigg[\sum_{p,q=1}^{M_2(i)} \frac{M_3(k, p) \cdot M_3(l, q)}{2} \left(\frac{c_{\text{config}}(3, \langle k, p \rangle) + c_{\text{config}}(3, \langle l, q \rangle)}{2} \right. \\
& + \left. \frac{c_{\text{config}}(2, k) + c_{\text{config}}(2, l)}{2} + c_{\text{config}}(1) \right) \Bigg] \Bigg)
\end{aligned}
$$

$$(4.2)$$

If the reconfiguration cost is identical within a layer ($c_{\text{config}}(i, \langle a, b \rangle) = c_{\text{config}}(i)$; i.e., a standard connector is used for different parts in the same layer) and all grippers and fingers have the same configuration sub-networks ($M_i(a, b) = M_i$; i.e., each gripper is able to mount the same number of finger types which have the same number of positions), Eq. (4.2) is simplified as follows.

$$
\text{Total reconfiguration cost: } C_{config} = \sum_{i=1}^{3} \left(\binom{M_i}{2} \left[\prod_{j=1}^{3-i} (M_j) \right] \left[\prod_{j=1}^{i-1} (M_{3-j}^2) \right] \sum_{j=1}^{i} c_{\text{config}}(4 - j) \right)
$$

$$(4.3)$$

Usually, each configuration is only good at handling a small range of objects. The relationship between R and the dimension ranges of the harvested targets is clear. A harvesting system has a relative larger R if it requires a high grasp quality for targets in a larger range.

4.1.5 Estimated Production Cost

The Production cost, ρ, is measured by the estimated load during a unit of time. The value of ρ is needed because it indicates the productivity of the system. More rapid operations result in higher production costs.

$$\text{Production cost:} \quad \rho = \frac{\lambda v}{N_{\text{arm}}\mu} \tag{4.4}$$

where

λ is the rate of the fruit density inside fields;
v is the velocity of a harvesting platform on trails;
μ is the mean service rate of harvesting processes by one arm; and
N_{arm} is the total number of arms.

To increase productivity, an automated harvesting system should have more arms (up to a limit due to congestion of arms), but the equipment cost will also increase. It is assumed that the production time (of picking a target and conveying it to a container) for each target is identically distributed at a rate of μ for any REE configuration, so ρ is independent of the configuration network and R.

4.1.6 Minimum Grasp Cost

The grasp cost reflects the grasp quality of picking fruits. The value of C_{grasp} is required because the grasp quality is the major factor affecting the quality of products by REEs in the harvesting processes. In each configuration, the grasp cost of a gripper is modeled as a function: $C_{\text{grasp}}(i, \delta)$, where d is the physical dimension of the grasped object.

The dimension distribution of targets is assumed to be known or estimated as $Pr(\delta)$. In a system with M configurations, the minimum grasp cost is derived as follows. The best configuration with minimum grasp cost is selected for each object based on its dimension.

$$\text{Minimum grasp cost:} \quad C_{\text{grasp}} = \int Pr(\delta) \cdot \min_{1 \leq i \leq M} \left(c_{\text{grasp}}(i, \delta)\right) \mathrm{d}\,\delta \tag{4.5}$$

where

C_{grasp} is the minimum grasp cost;
M is the total number of configurations ($M = M_1 \cdot M_2 \cdot M_3$); and
$c_{\text{grasp}}(i, \delta)$ is the grasp cost using configuration i.

4.1.7 Design Service Team for Reconfigurability

The total design cost of REE with multiple arms is estimated as J, the weighted sum of R, ρ and G (exact costs cannot be obtained due to task stochasticity).

$$\text{Total design cost:} \quad J = \beta_r N_{\text{arm}} \cdot C_{\text{config}} + \beta_\rho \rho + \beta_g C_{\text{grasp}} \tag{4.6}$$

where

N_{arm} is the number of arms that the harvesting system has;
C_{config} is the estimated reconfiguration cost;
P is the production cost;
C_{grasp} is the grasp cost; and
all β are normalization factors.
J is measured in cost units ($), and the design process aims to minimize the
   following optimization problem:

$$\min \quad J = \beta_r N_{arm} \sum_{i=1}^{3} \left[\binom{M_i}{2} \left[\prod_{j=1}^{3-i}(M_j) \right] \left[\prod_{j=1}^{i-1}(M_{3-j}^2) \right] \sum_{j=1}^{i} c_{config}(4-j) \right]$$
$$+ \beta_\rho \frac{\lambda v}{N_{arm}\mu} + \beta_g \int Pr(\delta) \min_{1 \le i \le \prod_{j=1}^{3} M_j} \left(c_{grasp}(i, \delta) \right) d\delta$$

$$s.t. \quad 1 \le N_{arm} \le \mathbb{N}_{arm}, \ 1 \le M_j \le \mathbb{M}_j, \ j = 1, 2, 3 \qquad (4.7)$$

This optimization is a nonlinear programming problem with integer decision variables N_{arm} and M_j. The values determine the best number of arms (N_{arm}) and the best configurations in each layer of configuration network (M_j). The objective function provides a numerical evaluation to the design of the reconfigurability in a REE system. By balancing the cost of reconfiguration and production workload with the quality of grasps, the number of arms and the configuration network are determined. The objective function captures the most important factors (quality and productivity) in designing reconfigurable systems. In Eq. (4.7), N_{arm} is the maximum number of arms available for the harvesting task. M_j is the maximum number of reconfiguration parts in j layer. Scalability analysis for the integrated automated platform is possible by adjusting the set of N_{arm} and M_j to different practical values, and then solving the optimization problem for each set.

Approximation methods of this nonlinear problem are left for future research (e.g., using genetic algorithms, or ant colony optimization, see Sect. 5.3.1). In the current work, we perform exhaustive search on Eq. (4.7) to determine the best configuration, because there are only limited number of REE systems under current research and development.

4.2   Centrality-Based Depot Allocation Protocol: Alignment to Enable Efficient Service Coverage

This section addresses research question 2: How to allocate the service team. In a resource-limited environment, the allocation policy of depots and the dynamic deployment of the service team (R) become critical. Using structural information of

Table 4.1 Popular centrality measures

Centrality measure	Definition	References
Betweenness centrality	The share of times that all shortest paths pass through the node being measured	[7]
Degree centrality	The number of links (or weighted number of links) connected to the node being measured	[8, 13]
Distance (closeness) centrality	The sum of the distances from the node being measured to all other nodes	[9, 15]
Eigenvector centrality	The principle eigenvector of the adjacency matrix defining the network	[2]

networks in responders' operations can improve the efficiency of disaster response [5]. In this work we develop a heuristic policy abstracted from the empirical study on the CPS disruption response problem [18]: Allocating depots based on node centrality in a client network.

Generally, node centrality reveals the importance of the node in the network, and several centrality measures are developed based on different structural information [3, 11]. Popular centrality measures and their definitions are summarized in Table 4.1.

In the current work, by using the centrality-based allocation (CBA), the depots are fixed to the nodes with the highest betweenness centrality in G. The betweenness centrality is selected because the nodes with high betweenness centrality are located on the important connection paths of the entire network. If service agents are located on these nodes, they can move quickly to other nodes and the service requests can be fulfilled in relatively short time. Besides, if the service requests are failures requiring responsive service, responders deployed at the nodes with high betweenness centrality can help to prevent failures from propagating to a large section of the network through the nodes on the critical paths.

$$\text{CBA policy:} \quad i = k, \; btw(k) \geq \max(btw(a)), \; i \in D, k \in N, a \in N \text{ and } a \notin D \tag{4.8}$$

$$i \neq j, \text{ for all } i, j \in D \tag{4.9}$$

where $btw(k)$ is the betweenness centrality of node k. After the depots are determined, agents are deployed to the depots according to the definition of depots (D) in Sect. 4.1.

4.2.1   Comparison with Random Depot Allocation

The CBA policy is hypothesized to be relatively more efficient than the random depot allocation (RDA): By deploying service agents to more "important" nodes, the agents can more quickly repair or prevent disruptions which otherwise can cause a large cascade of failures. RDA policy is defined as assigning each depot to a random node in a client network uniformly. It is an approximation of depot allocation without optimization. RDA policy has effective coverage of the nodes to handle random failures, but the cascading nature of failures in not considered. In this work, the hypothesis that CBA is a better depot allocation policy than RDA is tested. The hypothesis can be expressed as:

$$Hypothesis\ 1:\quad z(\text{CBA}) - z(\text{RDA}) < 0 \tag{4.10}$$

where z is the objective value indicating the total service latency and $z(a)$ is the objective value when using a depot allocation policy a.

4.3   Neuroplasticity-Inspired Scheduling Protocols: Response Operations to Minimize the Total Latency of Current and Emerging Tasks

This section addresses how to schedule the requested services. The N2N service is analogous to the lesion recovery process of human brain, which is a huge network of neurons.

Two important observations of neuroplasticity during the recovery process are reported in literature.

(1) Neurons try to maintain homeostatic state through activity-dependent development [4]. A neuron creates new spines and boutons (i.e., establish links) when its level of electrical activity is below a homeostatic set-point and decreases the number of links when its activity exceeds this set-point. In addition, neurons need a minimum level of activity to establish links.

(2) The lesion area of a brain will increase connectivity by random outgrowth of new connections during recovery [16]. By making the outgrowth, the local efficiency around an injured node is significantly increased, and it will not affect significantly the global efficiency of the entire brain network.

These two features of network plasticity, discovered from brain and neural research, provides insights of how artificially engineered networks can respond to failures.

4.3.1 Activity-Based Priority

In brain networks, the synapse reorganization after lesion can be explained by an activity-based wring model [4]. The activity of neurons will influence the speed of the growth and deletion of synapse (i.e., creation and rupture of links between neurons).

Initially after an injury, the injured brain region will quickly reduce its activity level to less than a minimum activation threshold (η) due to loss of external input. Thus, links are diminished. The peripheral regions of the lesion, however, will reduce its activity to between η and the homeostatic point (ε). Wiring additional links is enabled by the peripheral neurons which are eager to have more links. As more links are established between peripheral neurons and injured neurons, the entire brain network restores to normal operations. The recovery of brain lesion process is similar to the process of repairing a broken 2D network (e.g., a cast net). A net can only be fixed from the periphery to the center of broken area.

The same intuitive and efficient strategy can be applied in DLOC model. Service agents should try to restore links and nodes at the edge of a failed region. Otherwise, if some agents repair a component without any link to healthy nodes, the newly repaired node will fail again according to the cascading failure procedure.

Using an analogy, the connectivity (θ) in a client network can be used to represent the activity level as in the brain network. The speed of reconfiguration indicates the priority of allocating service agents. θ is the proportion of working links to the original degree as the node is working in normal operation. θ can be larger than 1 if auxiliary links are applied (see Sect. 4.4.2). The activity-based priority (ABP) model is illustrated in Fig. 4.3.

The priorities of service requested by nodes and links in the client network are expressed as in Eqs. (4.11) and (4.12).

$$\text{Node service priority:} \quad \omega(i, t) = \begin{cases} \theta(i, t) & \theta(i, t) < 1 \\ 2 - \theta(i, t) & \theta(i, t) \geq 1 \end{cases} \quad \text{for } i \in N, \ state(i, t) = 1 \quad (4.11)$$

$$\text{Link service priority:} \quad \omega(i, j, t) = \min(\vartheta(i, t), \vartheta(j, t)) \quad \text{for } (i, j) \in L, \ state\,(i, j, t) = 1 \quad (4.12)$$

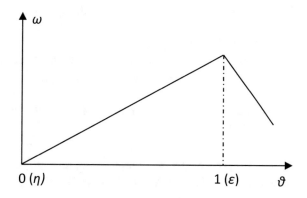

Fig. 4.3 Activity-based priority

where

t	is the time;
i, j	are nodes in the client network; and
(i, j)	is a link in the client network;
$\omega(i, t)$	is the priority to serve a request from node i at time t; and
$\omega(i, j, t)$	is the priority to serve a request by a link (i, j) at time t.

According to this model, a failed element which is located at the center of a failure area (with no links to living nodes), will receive less priority in the repair schedule than those failed elements, while still having connections to active nodes. The ABP in the DLOC model is expected to be relatively more efficient than the first-come-first-served (FCFS) protocol and the nearest neighbor policies.

4.3.2 Auxiliary Links

Besides activity-based development, synapse rewiring is observed in the brain, which associates functional and structural reconfiguration of the brain with information storage, learning, injury recovery, etc. [6, 12]. The random outgrowth of links in the injured area provides helpful modifications to surviving the entire brain networks. The random outgrowth is observed to follow the following policies [16]:

1. Significant increased local efficiency, indicated by the clustering coefficient of an injured node; and
2. No significant adjustment in the global efficiency, indicated by the average shortest path length of the entire network.

Inspired by the brain plasticity, the service team can apply a similar procedure to add auxiliary links (ALs) to the client network, to fulfill the N2N services. The mechanism of determining AL is shown as follows.

From the set of failures at the beginning of the cascading process (F_0), select a set of failed nodes which are not supervised.

$$\text{Target nodes:}\quad N_T \subseteq N, \quad \forall i \in N_T,\ \text{state}_0(i) = 1,\ \text{supervision}(i) = 0 \quad (4.13)$$

where N^{T} is the set of targeted nodes.

Then, find a node (j) in the second order neighbors of i. j is not supervised, and the degree of j is less than the mean degree ($deg_G(N)$).

$$\text{Target neighbor:}\quad j \in N_{(2)}(i), \quad i \in N_T,\ \text{supervision}(j) = 0,\ deg_G(j) < deg_G(N) \quad (4.14)$$

where

$N_{(2)}(i)$	is the set of the second order neighbors of node i.
$N_{(2)}(i)$	is defined as: $j \in N_{(2)}(i)$, $\exists l$ for (j, l) and $(l, i) \in L$; $(i, j) \notin L$.

AL Theorem *The clustering coefficient of i will be increased after adding an auxiliary link between i and j, if the number of links between j and i's neighbors ($l_{neighbor}(j, i)$) is larger than the product of the degree of i and clustering coefficient of i ($deg_G(i) \cdot \sigma_{cluster}(i)$).*

Proof If $k_i = 1$, $\sigma_{cluster}(i) = 0$. The new clustering coefficient $\sigma'_{cluster}(i) = 2 \cdot l_{neighbor}(j, i)/(2 * 1) > \sigma_{cluster}(i)$.

If $k_i > 1$, let e_i be the number links between the neighbors of node i.

$e_i = |\{(a, b): a, b \in N_{(1)}(i), (a, b) \in L\}|$

The original clustering coefficient is

$\sigma_{cluster}(i) = 2e_i/(deg_G(i) \, (deg_G(i) - 1))$

The clustering coefficient after reconfiguration is

$\sigma'_{cluster}(i) = 2(e_i + l_{neighbor}(j, i))/(deg_G(i) \, (deg_G(i) + 1))$

$\sigma'_{cluster}(i) - \sigma_{cluster}(i) = 2((deg_G(i) - 1) \, l_{neighbor}(j, i) - 2e_i)/(deg_G(i) \, (deg_G(i) + 1)(deg_G(i) - 1))$

If $l_{neighbor}(j, i) > deg_G(i) \cdot \sigma_{cluster}(i)$,

$\sigma'_{cluster}(i) - \sigma_{cluster}(i) > 0$.

End of proof.

Figure 4.4 illustrates an example of an auxiliary link. The red node (i) is failed at time 0. An auxiliary link changes the network's properties. The local clustering coefficient is increased for i (from 1/3 to 1/2).

By adding an auxiliary link at the beginning of the cascading process according to the AL theorem, the increased local efficiency will improve the robustness of the local area to cascading failures. At the same time, because the link added to the node has a lower degree than the mean degree, the AL will not significantly affect the overall connectivity of the entire network.

Note: The method of applying auxiliary links is only applicable to the client network with reconfigurability. Some cyber-physical infrastructures (CPI) are costly to create new links between nodes within a short time. The emerging trend in CPI design, however, shows that redundant and reconfigurable nodes and links will be created to assure enough active components [10]. For instance, transmission line switching

Fig. 4.4 Example of an auxiliary line for one failed node

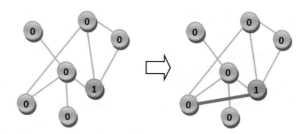

mechanisms are incorporated into system operator's post-disruption decision for a better mitigation effect [17].

4.4   Critical Performance Metrics: How Good Is the E-Work System in Terms of Time, Quality and Cost Metrics?

To analyze the control protocols over the DLOC model, three classes of metrics are designed based on the performance metrics in Table 3.2. They measure the most significant performance of the model for DLOC with different system parameters and control protocols.

4.4.1   Time Metrics

The total latency (z) is the most comprehensive measure, as it indicates the total service effectiveness. A lower z value implies the service team is more efficient in the N2N services (see the DLOC objective function, in Eq. (3.3)).

Average latency (ζ) indicates the average latency for each failure. It is often calculated only for the failures that are recovered.

$$\text{Average latency:} \quad \zeta = \frac{1}{|F|} \sum_{i \in F} (\sigma_i - \tau_i) \tag{4.15}$$

Another important indicator is the recovery time (t_{recover}), which shows the time the system is fully recovered from the initial failures.

$$\text{Recovery time:} \quad t_{\text{recover}} = \min\{t, |F_t| = 0\} \tag{4.16}$$

where F_t is the set of disrupted elements at time t, $F_t \subseteq F$.

4.4.2   Cost Metrics

The dynamic size of cascading failures ($|F_t|$) is also important, as it shows the process of cascading and service process. F_{max} is the largest set of failed elements at any time during the cascading process. It indicates the maximum influence of this disruption. The total number of failures ($|F|$) indicates the entire damage caused by the failures.

$$\text{Max cascade:} \quad |F_{\text{max}}| = \max_{t \geq 0} |F_t| \tag{4.17}$$

In a realistic problem, the total distance traveled by agents (Δ) is also an important indicator of effort and energy consumption in the response operation. The calculation of Δ is shown in Eq. (4.18).

$$\text{Total distance traveled by agents:} \quad \Delta = \sum_{i,j \in F \cup D} d(i, j) x_{ij} \qquad (4.18)$$

4.4.3 Quality Metrics

The quality of service (QoS) in the N2N problem has several perspectives. A direct measure of QoS could be the quality of the outcome. For example, in automated harvesting, the grasp quality (q) shows how well a target is detached by the robot arms and end-effectors. The number of serviced that can be handled in a unit of time shows the productivity of the service team. In automated harvesting, the total yield (Y) is the measure of productivity.

Preventability measures the impact of applied control mechanisms to disruptions. $P_{prevent}$ is calculated as the percentage of prevented disruptions that would be failed without response mechanisms, over the total disruptions without response mechanisms, as shown in Eq. (4.19).

$$\text{Preventability:} \quad P_{prevent} = \frac{\left|\tilde{F}\right| - |F|}{\left|\tilde{F}\right|} \qquad (4.19)$$

where $\tilde{F}$ is set of disruptions without response mechanisms ($R = \emptyset$).

Recoverability ($P_{recover}$) shows the resilience of the N2N service network. It indicates the probability of recovery after cascading disruptions.

$$\text{Recoverability:} \quad P_{reecover} = Pr(|F_t| = 0, t > 0) \qquad (4.20)$$

Other metrics can also be relevant when considering more complicated factors, e.g., resilience over time, learning and experience effects, training and quality of agents (level of intelligence designed in software agents).

References

1. Blanes C, Mellado M, Ortiz C, Valera A (2011) Technologies for robot grippers in pick and place operations for fresh fruits and vegetables. Span J Agric Res 9(4):1130–1141
2. Bonacich P (1972) Factoring and weighting approaches to status scores and clique identification. J Math Sociol 2:113–120

3. Borgatti SP (2005) Centrality and network flow. Soc Networks 27(1):55–71
4. Butz M, van Ooyen A (2013) A simple rule for dendritic spine and axonal bouton formation can account for cortical reorganization after focal retinal lesions. PLoS Comput Biol
   9(10):e1003259
5. Buzna L, Peters K, Ammoser H, Kuhnert C, Helbing D (2007) Efficient response in cascading
   disaster spreading. Phys Rev E 75:056107
6. Chklovskii DB, Mel BW, Svoboda K (2004) Cortical rewiring and information storage. Nature
   431(7010):782–788
7. Freeman LC (1977) A set of measures of centrality based on betweenness. Sociometry 40:35–41
8. Freeman LC (1978) Centrality in social networks conceptual clarification. Soc Networks
   1(3):215–239
9. Freeman J (1983) Emotional problems of the gifted child. J Child Psychol Psychiatry
   24(3):481–485
10. Krishna CM (2014) Fault-tolerant scheduling in homogeneous real-time systems. ACM
    Comput Surveys 46(4):1–34
11. Lee S (2012) The role of centrality in ambulance dispatching. Decis Support Syst 54(1):282–291
12. Murphy TH, Corbett D (2009) Plasticity during stroke recovery: from synapse to behaviour.
    Nat Rev Neurosci 10(12):861–872
13. Newman ME (2004) Analysis of weighted networks. Phys Rev E 70(5):056131
14. Rajan VN, Nof SY (1996) Cooperation requirement planning for multiprocessors: optimal
    assignment and execution planning. J Intell Rob Syst 15:419–435
15. Sabidussi G (1966) The centrality index of a graph. Psychometrika 31(4):581–603
16. Wang L, Yu C, Chen H et al (2010) Dynamic functional reorganization of the motor execution
    network after stroke. Brain 133(4):1224–1238
17. Zhao L, Zeng B (2013) Vulnerability analysis of power grids with line switching. IEEE Trans
    Power Syst 28(3):2727–2736
18. Zhong H, Nof SY (2014) DLOC complex network model for supply network disrup-
    tion response. In: Proceedings of international conference on production research—regional
    conference Europe, Africa and Middle East, Cluj-Napoca, Romania

Chapter 5
The TIE/DLOC Tool

5.1 Design of TIE/DLOC

The Teamwork Integration Evaluator (TIE) is a class of software tools that aims to evaluate the collaboration performance of e-Work. Since 1990s, different versions of TIE tools have been developed at the PRISM Center, Purdue University for evaluating various systems. The TIE tools apply parallel computers to simulate distributed e-Work enterprises, decision makers, agents, or sensors, which are communicating and collaborating in performing a set of tasks.

Agent-based modeling and simulation techniques are applied in TIE. Each agent is a collaborative unit in the e-Work. Agents make their own decisions in their e-Work environment to fulfill collaborative tasks. Tasks are generated randomly, based on the probability distributions observed in real-word. As agents can apply different protocols to collaborate with each other, the protocols can be evaluated and compared in different scenarios in the TIE simulator, before implementing them in real systems. All TIE tools focus on the effectiveness of allocating resources and agents to accomplish tasks, though they are implemented in different software languages. The common modules of all developed TIE versions so far are summarized in Table 5.1.

As the research on CCT has been evolving, the TIE tool becomes increasingly more elaborate, to model and evaluate more complicated interactions and collaborations. For example, the protocols shared by agents have been improved from the static assignment in earlier versions to dynamic administration (time-out and re-allocation) considering the failures during the execution of tasks under various conditions. Detailed comparisons of TIE versions are listed in Table 5.2.

Based on the foundation in previous versions, a new TIE tool, named TIE/DLOC, is developed for the current research. The major addition lies in the network models for clients and servers: The tasks requested by clients are interdependent based on

© Springer Nature Switzerland AG 2020
H. Zhong and S. Y. Nof, *Dynamic Lines of Collaboration*,
Automation, Collaboration, & E-Services 6,
https://doi.org/10.1007/978-3-030-34463-4_5

Table 5.1 Common modules in all TIE versions

Common module in TIE	Descriptions
Agent	Agent is a collaborative unit that performs tasks, makes its own decisions, and also communicates and collaborates with peers to accomplish the tasks
Teamwork protocol	A teamwork protocol is the decision-making mechanism for multiple agents in a distributed environment. The decisions on how to collaborate involve resource allocation, task assignment and planning, etc.
Task generator	Tasks are generated randomly according to certain probability distributions and task dependencies
Performance evaluator	Both general and task-specific evaluation metrics are implemented in each TIE version to measure the effectiveness of collaboration among the agents
Simulation controller	The simulation controller handles random numbers, simulation replications, and parallel executions of TIE

Table 5.2 Unique features in TIE Versions

TIE version	Major goals—performance evaluated	Unique features	References
TIE 1.1	Workflow integration and optimization	Coordination of distributed decision makers or designers in concurrent engineering systems	[2, 6]
TIE/Agent	Viability of each agent and of the entire system.	Design of agent-based manufacturing systems	[4]
TIE/Protocol	Task allocation ratio and profits of the servers	Solving the distributed resource allocation problem	[1]
TIE/MEMS	Efficiency of network communication models and architectures	Application of TIE to the design of wireless sensor networks	[5, 8]
TIE/TAP	Ratios for task allocation and task completion	Re-planning of tasks after failures in distributed task execution	[7]
TIE/DLOC	Time, cost, and quality of task completion	Network-to-network (N2N) with propagating services	(In this work)

the structure of the client network; and the servers are constrained by their collaboration ability. The new model and tool capture the increasing interdependency in collaborative e-Work systems. The input and output of TIE/DLOC are illustrated in Fig. 5.1.

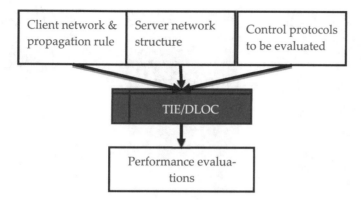

Fig. 5.1 Input-output diagram of TIE/DLOC

5.2   Software Architecture

The TIE/DLOC tool is developed with python and can run on Windows and Linux with parallel execution. The major dependent packages are *SimPy* [9] for discrete-event simulations and *NetworkX* [3] for network manipulations. The architecture of the TIE/DLOC is shown in the class diagram in Fig. 5.2.

TIE/DLOC has a modular structure and consists of eight classes of objects. The details descriptions of the classes are as follows.

Main
The main class handles the input and output of the TIE/DLOC tool. The input information includes:

1. the structure and size of a randomly generated client network, or a file containing the network structure of real systems;
2. the structure and size of the service team;
3. the propagation model of services (e.g., the cascading model of failures), and
4. simulation parameters (i.e., simulation length, number of replications, number of processors, and seed for pseudo-random number generation).

   The output information includes the mean and standard error of the mean, of all performance metrics as well as the number of failed elements at each time step.

Simulations
The simulation class consists of a set of functions to parallelize the experiments, to initiate each replication, to calculate the statistics of the experiments, and to format the results.

Strategies
The strategies class lists a set of functions to simulate different service strategies. As shown in Fig. 5.1., three protocols are modelled: The first come first served

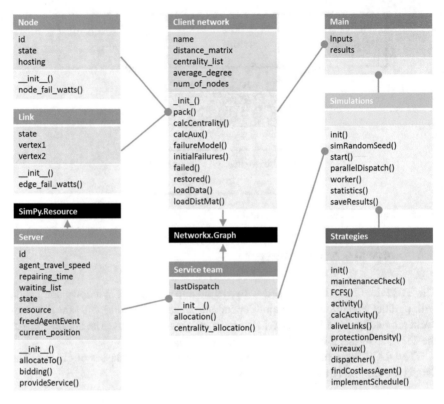

Fig. 5.2 Class diagram of TIE/DLOC

(FCFS) protocol, the activity-based priority, and the auxiliary lines inspired from neuroplasticity.

Client Network
The client network is the system which generates service requests. The class is used to generate the network or to load from existing models, and then calculate related parameters. Besides, the client network class also determines the service propagation model (i.e., the failure model), and inserts the initial service (i.e., failure) into the network.

Service Team
The service team class creates the network of service agents. In the current tool development, regular network with fixed degrees is developed. Different network structure will be experiment in future work. The service team class also assumes the function of allocating the depots of service agents to the client network. Random depot allocation and betweenness-centrality based allocation are implemented in this TIE/DLOC model.

Node

The node class represent an element in the client network. The class keeps a record of the state of this element: working or failed. To model the propagation of services, the propagation mechanism is coded as a method in the node class.

Link

The link class is used to generate the objects showing the dependences between nodes. Besides showing the state of the link (working or failed), the class also shows the attributes of the two vertexes of the link. To model the propagation of services, the propagation mechanism is coded as a method in the link class.

Server

The server class is used to model a single agent. It has attributes of the speed of traveling, the estimated service time, the list of clients to be served, the state of the agent (busy or idle), the resource attached to the agent, and the current position of this agent. Each server has the function to travel (or be allocated) to a destination, the function to bid for service requests, and the function to provide services.

References

1. Anussornnitisarn P, Nof SY, Etzion O (2005) Decentralized control of cooperative and autonomous agents for solving the distributed resource allocation problem. Int J Prod Econ 98(2):114–128
2. Ceroni JA, Nof SY (2002) A workflow model based on parallelism for distributed organizations. J Intell Manuf 13(6):439–461
3. Hagberg AA, Schult DA, Swart PJ (2008) Exploring network structure, dynamics, and function using NetworkX. In: Varoquaux G, Vaught T, Millman J (eds) Proceedings of the 7th python in science conference (SciPy2008), Pasadena, CA USA, pp 11–15
4. Huang CY, Nof SY (2002) Evaluation of agent-based manufacturing systems based on a parallel simulator. Comput Ind Eng 43(3):529–552
5. Jeong W, Nof SY (2008) Performance evaluation of wireless sensor network protocols for industrial applications. J Intell Manuf 19(3):335–345
6. Khanna N, Nof S Y (1994) TIE: teamwork integration evaluation simulator, a preliminary user manual for TIE 1.1. School of Industrial Engineering, Purdue University, PRISM Center
7. Ko HS (2010) Design of protocols for task administration in collaborative e-work systems. Ph.D. dissertation, School of Industrial Engineering, Purdue University
8. Liu Y, Nof SY (2004) Distributed microflow sensor arrays and networks: design of architectures and communication protocols. Int J Prod Res 42(15):3101–3115
9. Muller K (2004) Advanced systems simulation capabilities in SimPy, presented at Europython 2004

Chapter 6
Applications and Experiments

6.1 Precision Agriculture: ACRP DLOC Method and Protocol in the Design of Reconfigurable End-Effectors (REE)

The first set of experiments is designed to test the effectiveness of ACRP, applied to determine the configurations of the reconfigurable end-effectors (REE) of a mobile harvesting platform (described above, and detailed in Chap. 4). The platform is a mobile robot which can carry multiple arms. The arms and their end-effectors are subject to uncertain disruptions, in terms of the unknown properties of individual fruits to be harvested next.

Each arm can hold a REE which has up to three layers of reconfiguration levels. To design the platform configurations, the objective function of the total design cost (Eq. (4.7)) should be minimized. In the current work, several REE models and their possible combinations are evaluated. To validate the design results, the designed harvesting systems are simulated by TIE/DLOC. The performance measures of the alternative systems are obtained and compared.

6.1.1 Design of the REE Experiments

An automated harvesting system is simulated by TIE/DLOC tools. The major experimental parameters used are listed in Table 6.1.

The rate of fruit (λ) is obtained from field study [6]. The fruits (e.g., tomatoes and pumpkins) are modeled as balls with different radius (r: uniform distributed $U(10, 100$ mm$)$). The largest fruits are 10 times larger than the smallest ones. This situation is observed in *companion cropping* aimed at reducing pests and diseases [9]. Whenever the prediction of future fruits is necessary during the simulated harvesting

© Springer Nature Switzerland AG 2020
H. Zhong and S. Y. Nof, *Dynamic Lines of Collaboration*,
Automation, Collaboration, & E-Services 6,
https://doi.org/10.1007/978-3-030-34463-4_6

Table 6.1 Parameters and assumptions for design of REE experiments

Parameter	Description	Value and reference
λ	Mean rate of fruit density, following an exponential distribution	83 1/m (along a trail in fields) [6]
v	Harvesting platform forward velocity	7.2 m/h [5]
μ	Mean harvesting service rate, following an exponential distribution	1/11 1/s [5]
$c_{config}(1)$	Reconfiguration cost in the first layer	0.9 ($)
$c_{config}(2)$	Reconfiguration cost in the second layer	0.8 ($)
$c_{config}(3)$	Reconfiguration cost in the third layer	0.1 ($)
$Pr(\delta)$	Distribution of fruit dimensions	Uniform distributed: U(10, 100) mm
$c_{grasp}(i, \delta)$	Grasp cost	See Fig. 6.1
n_R	Number of replications for each scenario	100
t_{max}	Simulation time-duration for each scenario run	1 h
$|\Psi|$	Number of scenarios	7 (see Table 6.3)

process, the average value of $Pr(\delta)$ for the most recent harvested items is used for the future jobs.

The harvesting platform parameters (platform velocity: v, and harvesting service rate: μ) are obtained from a previous prototype robot for harvesting [5].

The reconfiguration costs ($c_{config}(i)$, $i = 1, 2, 3$) are estimated based on the difficulty and time of changing structures in each given layer. To change the entire gripper ($c_{config}(1)$) is considered to be relatively more difficult than to change a finger ($c_{config}(2)$) and thus more costly than to change a finger posture ($c_{config}(3)$). As shown in Table 5.1, $c_{config}(1) > c_{config}(2) > c_{config}(3)$.

Grasp quality is calculated by using *Graspit!* [8], which is a simulation tool developed for automated grasp research. Grasp quality is a force closure index obtained from *Graspit!*. Within the range of fruit sizes defined in Table 6.1, the higher the value of the index is, the better the target is grasped. The grasp quality distributions of three example grippers over the fruit sizes are shown in Fig. 6.1. The grasp cost ($c_{grasp}(j, \delta)$) is assumed to be one minus the grasp quality. Table 6.2 provides examples of a gripper and shears used in the harvesting experiments, as they are commonly used in such cases.

Seven scenarios (Ψ) are simulated in this experiment. Each scenario has 100 replication runs, and each replication time-duration lasts 1 h.

6.1.2 Results of REE Experiments

Seven scenarios in the design phase were tested, representing seven different configuration networks. The detailed structures of these scenarios are shown in Table 6.3.

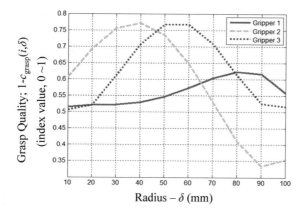

Fig. 6.1 Optimal grasp quality ($1 - c_{grasp}(i, \delta)$) of three grippers as functions of fruit, radius range (10, 100) mm (computed by using *Graspit!* [8])

Table 6.2 Example tools in the automated harvesting experiments

Picture	Function	Example grasp
	Pick fruits	
	Prune branches	

Values in the cells of Table 6.3 represent the numbers of configuration options in the corresponding layers of configuration networks. Those REE system scenarios are chosen according to several reconfigurable models in previous relevant research, and their potential combinations.

Table 6.3 Experimental scenarios tested for REE

Ψ_i	OOR	No. of arms (N_{arm})	Parts in the 1st layer (M_1)	Parts in the 2nd layer (M_2)	Parts in the 3rd layer (M_3)
1	1	1	2	1	1
2	3	1	4	3	2
3	1	2	2	1	1
4	2	2	2	3	1
5	1	2	4	1	1
6	3	2	4	3	2
7	1	3	4	1	1

Edan and Miles [5] tested harvesting with up to six arms. Each of the arms had an independent gripper. In the REE case, limited interchangeable parts are available, hence, up to three arms are tested. Two types of fingers with different contact types (line and point) and two finger postures were tested by Yeung and Mills [16]. In the current experiment, up to two postures and up to three types of fingers are tested, for up to four different end-effectors.

The tested models are mapped into the configuration networks: In the first layer, the node represents shears or grippers; different types of fingers are in the second layer; and the third layer represents finger positions for each type of fingers. In the simulation, it is assumed that all fingers can be mounted on all grippers. For the grippers, it is assumed that all the grippers have the same mechanisms for reconfiguration, but each gripper configuration produces a different grasp quality.

After calculating for each of the seven scenarios its value of the objective function (Eq. 4.7), the total design cost (J) for each of the seven scenarios was computed, and it is shown in Fig. 6.2. From the observations given in Fig. 6.2, under the stated experimental conditions, the automated harvesting system design with the lowest value of J, scenario 7, is equipped with three arms ($N_{arm} = 3$) and four tools ($M_1 = 4$). Each end-effector has one set of reconfigurable parts ($M_2 = 1$) and one finger posture ($M_3 = 1$). In scenario 7, only layer 1 has multiple reconfiguration possibilities,

Fig. 6.2 Objective function (J) values of seven simulated REE configurations

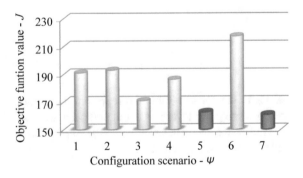

hence, the OOR of this system is one, by definition. (Scenario 5 had the next lowest value of J; see the observations discusses below, at the end of this section.)

6.1.3   Validation with Automated Harvesting Simulations

The result of the ACRP design phase is validated through testing the seven scenarios in mathematical simulation built on TIE/DLOC. The detailed settings for all seven scenarios are as follows; the simulation results are shown in Figs. 6.3 and 6.4.

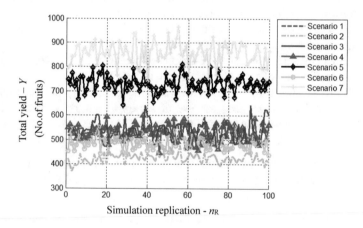

Fig. 6.3 Total yield (Y) with seven REE configuration scenarios

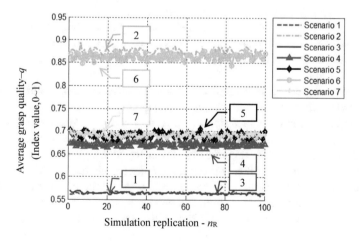

Fig. 6.4 Average grasp quality (q) in different scenarios

1. The automated mobile platform carries a tool rack and one arm (scenarios 1–2), two arms (scenarios 3–6), or three arms (scenario 7).
2. The reconfiguration process is modeled by a sequence of events triggered by the detection of new items ("disruptions"). If a reconfiguration task does not require changes of grippers or fingers, the arm will keep them (instead of returning them to the tool rack and getting them back again). If a reconfiguration is requested, the arm first dismounts its fingers and gripper, and sends them to the tool rack. Then, the new configuration parts will be fetched from the tool rack. If a part is currently unavailable, the arm will wait until another arm releases the necessary resource.
3. In each scenario, the automated harvesting system uses an optimization algorithm to determine which configuration should be used for each target [18].
4. In each experiment, the harvesting system is assigned to collect fruits continuously, along a field trail, within 100 min.
5. Fruits are detected by the mobile platform as a sequence modeled to be a Poisson process, with a rate of six fruits per minute ($\lambda \cdot v$).
6. Besides harvesting fruits, an arm will have 1% probability of encountering a bypass situation that requires the use of shears to remove a branch. (1% is assumed as a relatively low probability, compared to practice, to enable testing the significance of difference performance, even under such minimalist assumption.)

100 replications of the simulations were performed. The total yield (Y), and the average grasp quality (q) for each scenario are shown in Figs. 6.3 and 6.4. Several observations can be made, as follows.

1. Scenarios 7 and 5 had relatively higher yield according to the simulation (Fig. 6.3).
2. Although scenarios 2 and 6 resulted in the best grasp quality (Fig. 6.4), their configuration structures were more complicated: Both of them had 3OOR, as shown in Table 6.3. As a result of their relatively more complicated configuration network, the automated harvesting system in scenarios 2 and 6 had to cost relatively more time to reconfigure the end-effectors, in order to achieve the best grasp quality.
3. Scenarios 1 and 3, with relatively fewer end-effector configuration options, yielded significantly inferior average grasp quality (Fig. 6.4).
4. Scenarios 4, 5, and 7 yielded similar q (Fig. 6.4). Scenario 4, however, had 2OOR compared with 1OOR for scenarios 5 and 7, thus its production yield in terms of total fruits harvested was significantly lower (by 33%).
5. Considering the trade-off between grasp quality (q) and the total yield (Y), automated harvesting systems with scenarios 7 or 5 are the best for handling and control of disruptions in the simulated harvesting task. It is also consistent with the optimization result obtained in the ACRP framework (see Fig. 6.2).

6.2   Critical Infrastructure Protection: DLOC Protocols and Policies for Collaborative Disruption Response (CDR)

To validate the DLOC model and test the effectiveness of the control protocols, experiments are conducted on different client networks. The experiments are designed to test whether DLOC is an efficient modeling tool that can:

- characterize effectively the cascading of failures in large-scale CPIs; and
- model and analyze the collaboration of agents in optimizing their response to the disruptions.

Sections 6.2.1–6.2.4 report on experiments for validating the DLOC model and testing the developed depot allocation polices; and Sect. 6.2.5 details the experiments on testing the developed scheduling protocols for service agents.

6.2.1   Design of CDR Experiments on Different Depot Allocation Policies

Complex network models (ER, WS, and BA) can approximate current and emerging CPIs with certain accuracies [3, 4, 12, 15]. In the current experiments, the behavior of DLOC is examined over these three network models assuming the same number of nodes (500). To design experiments with networks that are comparable, their generating parameters are chosen such that the Mean degree of each is 4. The degree distributions of simulated network models are shown in Fig. 6.5.

At time 0, it is also assumed that a CPI has five random failed nodes.

The cascading threshold φ is set to 0.25.

Fig. 6.5 Degree distributions of three tested CPI network models

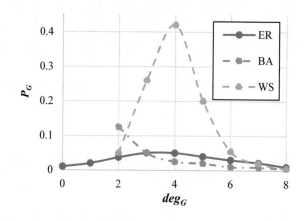

The CPI client network is interacting with a response team (R), the disruption handling providers network. In the experiments, a fixed degree is used for each agent to form a regular graph for R.

Agents' travel speed is assumed to be one link per time unit ($v = 1$ link/h) and it requires a single time unit to repair each failed node or link ($t_{\text{repair}} = 1$ h).

When the agents try to protect the system from collapse, two critical parameters affect the response performance:

- Total number of agents (n_A); and
- Their degree of connection ($deg_R(A)$) in the response team.

To validate the DLOC model and to test the depot allocation policies, the same first-come-first-served (FCFS) scheduling rule is used in different CPI models with different depot allocation policies. (Improving the response operation itself is planned in future work.) The effectiveness of FCFS has been validated with experiments [17]. The protocol assigns the closest available agent to the disrupted (broken or disabled) element according to time sequence. A summary of the protocol is given as follows.

FCFS scheduling rule

1. DEFINE an array as failure list: F
2. LOOP at each time step:
3.   IF |F| > 0:
4.   POP the first element of F as i.
5.   IF γ_i = Ø (i is a node failure):
6.   Find an agent α in A so that the time period from the current time to when i finishes its task + α travels from its last connection to u_i node is minimum.
7.   Assign state(α) = 1 (after the repair, state(a) is back to 0)
8.   Push i to a's waiting queue; update a's task finishing time and last connection.
9.   IF γ_i = j ((i, j) is a link rupture):
10.   For i, find a according to steps 6-8.
11.   For j, find b according to steps 6-8, but only from agents who are able to collaborate with a.
12.   Detect new failures in G and append them to F.

Three sets of experiments are performed as reported in the following three sections.

6.2.2 Experiments on Different Models of Cyber-Physical Infrastructures

In the first set of experiments, the performance of different depot allocation models on three CPI models are tested. The experimental parameters are summarized in

Table 6.4 Parameters for the design of experiments on different CPI models

Parameter	Description	Value
n_A	Number of agents	35
$deg_R(A)$	Degree for each agent	4
v	Agents' traveling speed	1 link/h
t_{repair}	Repair timespan	1 h
$|N|$	Number of nodes in CPI models	500
$\overline{deg_G}$	Mean degree in CPI models	4
ϕ	Cascading threshold	0.25
u	Spread speed of failures (disruptions)	1 link/h
$|F_0|$	Number of initial failures (disruptions)	5
n_R	Number of replications for each scenario	100
t_{max}	Simulation duration for each scenario	40 h
$|\Psi|$	Number of scenarios	9 (3 allocation policies $\times$ 3 CPI network models)

Table 6.4. The response team has 35 agents and $deg_R(A) = 4$. Figure 6.6 shows the failed nodes during the initial 40 h, with depot allocation policies CBA and RDA and without disruption remediation ($R = \emptyset$). Each simulation setting has 100 replications.

The following observations can be made based on the results shown in Fig. 6.6.

- According to the degree distribution (Fig. 6.6a), the ER random graph has scattered components, so random failures (disruptions) will have less chance to influence other components. Hence, the ER random graph is relatively more resistant to cascading effects than other models in the experiments.
- The small-world model is affected relatively more than the other two networks, because the failures are easier to propagate when most of the nodes are connected within a small number of steps (i.e., the small-world phenomenon).
- The response behaviors are affected by the network structures:

  1. For the random graph, the resources are more than adequate to diminish the failures (Fig. 6.6a).
  2. For the scale-free network, as the response resources are limited, the depot allocation policy becomes an important factor. A response using CBA can successfully restore the entire network, but with RDA the response cannot (Fig. 6.6b).
  3. For the small-world model, the resources are far from sufficient, tuning the depot allocation policy makes no difference on the response behavior. Neither CBA nor RDA can help to restore the CPI from failures (Fig. 6.6c).
  4. Networks' behavior with no disruption response ($R = \emptyset$) is shown in Fig. 6.6a–c to illustrate the preventability (P) of these networks. The response team can

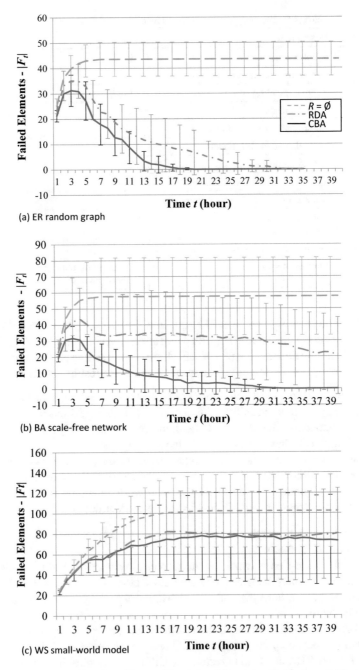

(a) ER random graph

(b) BA scale-free network

(c) WS small-world model

Fig. 6.6 Failed (disrupted) nodes with $n_A = 35$ and $deg_R(A) = 4$ (vertical bars are 0.95 confidence internals)

Table 6.5   Experiment results ($n_A = 35$ and $deg_R(A) = 4$)—Total Latency (z; h)

CPI model	RDA		CBA		P-value of two-sample t-test*
	Avg.	STD.	Avg.	STD.	
ER	573.3	569.4	504.1	588.8	0.123
BA	864.5	995.1	337.2	489.3	0.000
WS	1623.2	942.9	1462.0	967.6	0.029

*The last two tests on equal means are rejected at significance level 0.05

prevent more disruptions in the ER model than in the BA model and in the WS model.

The experiment results can be interpreted through the total response time (z) as the fundamental objective of disruption response. The network structures of CPIs determine the response performance when the resources are identical (see Table 6.5), as follows.

1. In ER random graph, the CBA makes no significant improvement on z, as the resource are abundant.
2. In BA scale-free network, the z is reduced by 61% by CBA.
3. In WS model, as relative scarce resources are supplied, z is only slightly (less than 10%) improved by CBA, compared with RDA.

6.2.3   Experiments with Different Service Team Size

In order to further test the difference between RDA and CBA, a second set of experiments has been conducted with different n_A and $deg_R(A)$ on the same set of scale-free networks. The experimental parameters are listed in Table 6.6.

Table 6.7 and Fig. 6.7 show the results measured for important metrics: Total response time (z), Max cascade ($|F_{max}|$), Total distance traveled by agents (Δ), and Preventability (P). The following observations can be made:

- As shown in Table 6.7, $|F_{max}|$, z, Δ, and P are all improved by the CBA policy, within the given scale range of the response team ($30 \leq n_A \leq 42$; $1 \leq deg_R(A) \leq 28$). Thus, *Hypothesis* 1 is valid in the critical range.
- Similar results can be obtained in the ER random graph and WS small world model in ranges $26 \leq n_A \leq 28$ and $56 \leq n_A \leq 60$, respectively. Outside these critical ranges, the response resources are either too abundant, or too scarce, and the depot allocation policy does not influence the performance significantly.
- The role of collaboration is also important in disruption response. As shown in Fig. 6.7b, when n_A is fixed, increasing the node degree ($deg_R(A)$) in the response team reduces the total response time in both CBA and RDA. Thus, in a limited resource environment, increasing the collaboration ability among agents (by, for example, effective training, workflow optimization, advanced tele-collaboration

Table 6.6 Parameters for the design of experiments with different service team size

Parameter	Description	Value
n_A	Number of agents	$30 \leq n_A \leq 42$
$deg_R(A)$	Degree for each agent	$1 \leq deg_R(A) \leq 28$
v	Agents' traveling speed	1 link/h
t_{repair}	Repair timespan	1 h
$|N|$	Number of nodes in CPI models	500
u	Spread speed of failure (disruption)	1 link/h
$\overline{deg_G}$	Mean degree in CPI models	4
ϕ	Cascading threshold	0.25
$|F_0|$	Number of initial failures (disruptions)	5
n_R	Number of replications for each scenario	100
t_{max}	Simulation duration for each scenario	40 h
$|\Psi|$	Number of scenarios	108 (2 allocation policies × 54 service teams)

Table 6.7 Summary of results from the experiments with different team size (see Fig. 6.7)

Metric	Average percentage increase from RDA to CBA (%)[a]	Standard deviation of the percentage increase	p value of one-sample t-test**
Max cascade ($|F_{max}|$)	−29.43	0.0291	0.000
Total distances traveled by agents (Δ)	−45.85	0.1013	0.000
Total response time (z)	−61.52	0.1126	0.000
Preventability (P)	244.53	0.6777	0.000

[a]The percentage increase is calculated as follows: Subtract the metric of RDA from the metric of CBA, and then divide by the metric of RDA
**The p values show all hypotheses of no increase are rejected at significance level 0.05

technologies) is important to allow agents to establish more lines of collaboration with peers, and consequently enhances the disruption response performance.

6.2.4   Case Study: Disruptions in a Water Distribution Network

The experiments above are all deterministic. In real life, the traveling time of agents (t_D), the repair timespan (t_{repair}) and the failure cascading process (the *CASCADE*

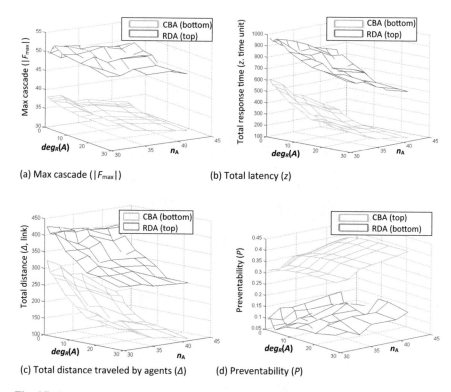

Fig. 6.7 Results of experiments with different n_A and $deg_R(A)$ on scale-free CPI models

function) all involve uncertainty. In this section, a disruption handling and control case study in the Hetch Hetchy water system (HH) is presented to illustrate and validate how the DLOC model and the depot allocation policies work in a stochastic scenario.

Smart water distribution network (WDN) is an emerging CPI that can provide better monitoring and controls for urban (and non-urban) water supply. It is reported that cascading effects can disrupt a WDN if a single pump station or pipeline is destroyed by natural disaster, intentional damage, or harmful software [11]. Concurrent collaboration is often mandatory in repairing a leak. The process involves detecting the leak, turning off the correct pipe, bringing in alternative water supply, repairing and restoring the physical location, etc. [13]. The steps require concurrent collaboration by a team of agents at different locations.

HH is a real, complex system. It serves 2.6 million residential, commercial, and industrial customers in San Francisco bay area from 11 reservoirs [1]. Besides water, the hydraulic system also supplies electricity to the city through three powerhouses. A schematic map of the HH augmented with CPI nodes and agents is shown in Fig. 6.8. Each node in the system is a reservoir, powerhouse, water station, treatment facility, pump station, or valve house. The links between the nodes are pipelines,

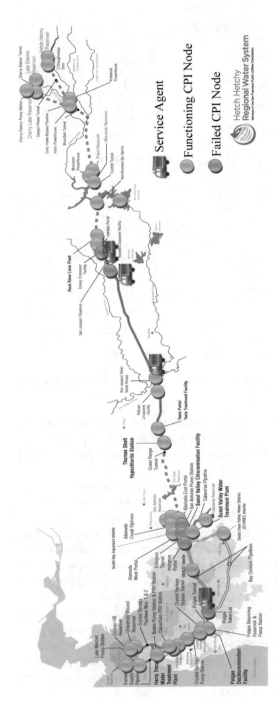

Fig. 6.8 HH network representation and service agents. *Map source* SFPUC [10]

tunnels, rivers, power transmission lines and cyber cables for communication. The entire network is connected and has 48 nodes. The average degree is 3.4583. The highest betweenness centrality node is observed at the Alameda Creek Siphons. Other nodes with high betweenness centrality are also found around the Sunol Valley Chloramination Facility, which is at the center of the entire system connecting the water sources and the service areas.

A graphical variation of TIE/DLOC is coded with *AnyLogic* software, which is the environment for the case study. Different from the conceptual model in the previous sections, the HH network is weighted by the distance an agent needs to travel from one node to another.

Failures in the HH are assumed to follow a similar cascade procedure as described in Sect. 3.3. The cascading threshold φ for each node is randomly and uniformly selected from $U[0.2, 0.25]$.

The response team for the case study is assumed to have fixed degrees, $deg_R(A)$.

Agents' travel speed is set to 10 miles per hour and the distance between two nodes in the straight-line distance between them plus a random 10% compensate. The movement seems slow, but is a realistic estimate considering the difficulty in accessing the mountainous area and the traffic congestion due to infrastructure damages. The repair time follows a uniform distribution ($t_{repair} = U(1, 3)$ (h)). The experimental parameters are summarized in Table 6.8.

To characterize the DLOC model, the two depot allocation policies discussed in Sect. 3.3 are compared in this case study. The agent models for the two policies have the same HH layout and the same random events enforced by the same seeds for generating pseudorandom numbers. An initial failure is tested at each node in HH, and each test has 10 replications.

Figure 6.9 illustrates the comparisons of RDA and CBA on failed nodes and links after 50 h ($|F_{50h}|$), max cascade $|F_{max}|$, total travel distance by agents (Δ),

Table 6.8 Parameters for the case study in HH water distribution network

Parameter	Description	Value
n_A	Number of agents	$12 \leq n_A \leq 18$
$deg_R(A)$	Degree for each agent	$2 \leq deg_R(A) \leq 8$
v	Agents' traveling speed	$10 + U(-1,1)$ (miles/h)
t_{repair}	Repair timespan	$U(1, 3)$ (h)
G	Client network	HH water distribution network
ϕ	Cascading threshold	uniform(0.2, 0.25)
u	Spread speed of failures (disruptions)	1 link/h
$|F_0|$	Number of initial failures (disruptions)	1
n_R	Number of replications for each scenario	10
tmax	Simulation duration for each scenario	40 h
$|\Psi|$	Number of scenarios	4704 (2 allocation policies × 49 service teams × 48 initial failures)

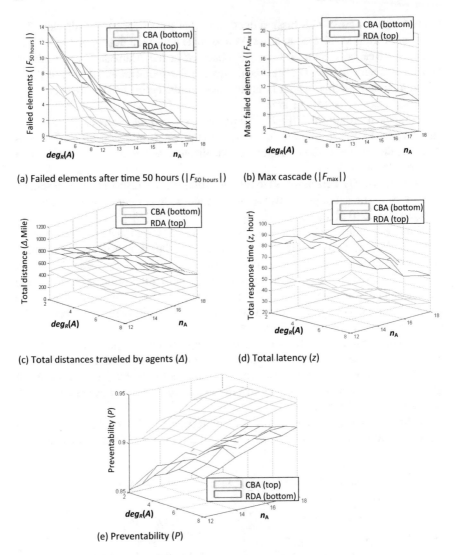

(a) Failed elements after time 50 hours ($|F_{50\,hours}|$)   (b) Max cascade ($|F_{max}|$)

(c) Total distances traveled by agents (Δ)   (d) Total latency (z)

(e) Preventability (P)

Fig. 6.9 Results of HH case study for five performance metrics

and total disruption response time (z). The tested range of the response team size is $12 \leq n_A \leq 18$, and $2 \leq deg_R(A) \leq 8$. Those measurements show the efficiency of a disruption response network fulfilling the requirements of response earliness, response effectiveness, and reliability in preventing cascading effects.

Based on the results measured in the experiments, as shown in Fig. 6.9, the following observations can be made:

- the CBA policy, Fig. 6.9a, successfully limits the spreading of failures in the HH network within the tested range of the response team.

- After 50 h, the remaining failed elements are 60% less in CBA than RDA, on average.
- Measured metrics are shown in Fig. 6.10b–e, and the improvements relative to RDA to CBA are calculated and tested (see Table 6.9). The significant improvement can be explained, because by using CBA, the agents are prepared at more important nodes in the CPI network.
- The importance of collaboration in the CPI disruption response is observed in Fig. 6.9. With the increase of collaboration of agents, the performance measurements are all improved significantly, for both CBA and RDA. When agents are able to collaborate with more peers, they can have better response performance.

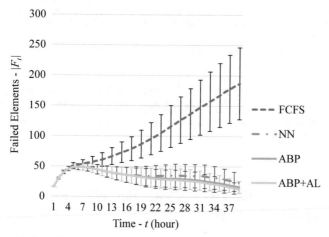

(a) Overall failed elements in experiments with four scheduling protocols

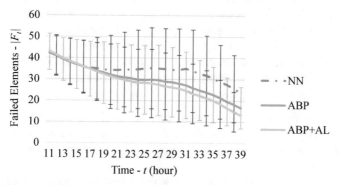

(b) Detailed comparisons from 11 to 40 hours after initial failures

Fig. 6.10 Number of failed elements after initial failures (vertical bars are 0.95 confidence intervals)

Table 6.9 Summary of results for the case study in HH water distribution network (see Fig. 6.9)

Metric	Average percentage increase from RDA to CBA (%)[a]	Standard deviation of the percentage increase	p value of one-sample t-test**
Max cascade ($|F_{max}|$)	−34.13	0.0588	0.000
Total distances traveled by agents (Δ)	−49.25	0.0706	0.000
Total response time (z)	−47.50	0.0547	0.000
Preventability (P)	4.06	0.0102	0.000

[a]The percentage increase is calculated as follows: Subtract the metric of RDA from the metric of CBA, and then divide by the metric of RDA
**The p values show that all hypotheses of no increase are rejected at significance level 0.05

In general, from this case study, it is observed that: With the same settings of system and network structure, just by changing depot allocation policy and adding collaboration capability of the agents, CDR performance can be significantly improved within the tested range.

6.2.5 Experiments with DLOC Neuroplasticity-Inspired Scheduling Protocols

To test the significance of the developed DLOC neuroplasticity-inspired response protocols, a power grid system and its disruption response operations are simulated with TIE/DLOC. The client network model is built based on the real power grid of western United States [14]. Collectively, there are 4941 nodes in the network, representing generators, transformers, and substations. 6594 links between the nodes represent the high voltage transmission lines between the nodes.

For the experiments, we have made the following assumptions.

- To represent disruptions, five initial failures are randomly injected to the network. A failure is assumed to propagate over the network according to the Watts cascading failure model. The propagation speed is one element per hour. The failure threshold is 0.25.
- Agents form a service team to respond to the failures over the power grid network. The agents will travel 1 link per hour, and the repair operations will take 1 h for each failure.
- The betweenness centrality-based depot allocation methods are used in this experiment. The simulations run 100 replications, and each replication duration is 40 h.

Table 6.10 Parameters of the design of experiments with neuroplasticity-inspired scheduling protocols

Parameter	Description	Value
n_A	Number of agents	500
$deg_R(A)$	Degree of each agent	300
n_R	Number of replications	100
G	Client network	Power grid of western United States
$|F_0|$	Number of initial failures (disruptions)	5
u	Spread speed of failures (disruptions)	1 element per hour
ϕ	Cascading threshold	0.25
v	The travel speed of agents	1 link per hour
t_{repair}	Repair timespan	1 h
n_R	Number of replications for each scenario	100
t_{max}	Simulation duration	40 h
$|\Psi|$	Number of scenarios	4 (4 scheduling protocols)

- Four different control protocols of the response operations are compared in the experiments:

  1. The efficacy of the developed neuroplasticity-inspired scheduling protocols, i.e., activity-based priority (ABP), and ABP plus auxiliary links (AL), are compared with the first-come-first-served (FCFS) and the nearest neighbor (NN) policies.
  2. Unlike the FCFS, which is not an optimized scheduling protocol, the NN policy often yields good results in literature [2, 7]. Instead of arranging the services according to the sequence of requests, servers should respond to the closest available request after every service completion.

- In this experiment, the service team has 500 agents, and each of them is able to collaborate with 300 peers. The experimental parameters are summarized in Table 6.10.

  After running the simulations with TIE/DLOC, the results, performance metrics, and observations based on them are as follows.

- The processes of cascading failures with response controls after the initial failures are shown in Fig. 6.10.
- Figure 6.10a shows the overall processes, and Fig. 6.10b illustrates the detailed comparisons of NN, ABP and ABP + AL from the 11th hours after the initial failures (disruptions). The graphs clearly show that the control protocols of ABP and ABP + AL improve the response performance by slowing down the cascading process of failures.
- In each experiment, the failure processes with different protocols are similar during the first five hours. It is because for a failure occurring at a random location, there

Table 6.11 Summary of the results of experiment with DLOC neuroplasticity-inspired scheduling protocols

Scheduling protocol	FCFS		NN		ABP		ABP + AL	
Statistics	Avg.	SEM	Avg.	SEM	Avg.	SEM	Avg.	SEM
Preventability ($P_{prevent}$)	0.482	0.010	0.575	0.006	0.559	0.005	0.561	0.005
Total distance traveled by agents (Δ, number of links)	1244.34	124.29	1507.02	232.85	1426.88	211.00	1405.95	210.67
Total failures ($|F|$)	488.58	56.11	179.62	27.90	176.25	24.91	169.28	23.89
Average latency (ζ, hour)	6.974	0.029	7.501	0.038	6.954	0.038	7.028	0.038
Total latency (z, hour)	3984.38	535.67	1406.46	245.62	1266.05	204.92	1222.59	199.73
Recoverability ($P_{recover}$)	0.77		0.99		1.0		1.0	
Recovery time (considering only recovered cases, recover, hour)	89.12	31.42	23.44	1.63	25.28	2.34	23.59	1.65

are no response agents deployed at near nodes to immediately handle the failures. After the first five hours, the difference between the FCFS and other protocols becomes significant.

- The statistics and comparison of performance are shown in Tables 6.11 and 6.12. By applying the DLOC neuroplasticity-inspired scheduling protocols, the percentage of recoverable replications is increased from 77% to 100%, compared with the FCFS protocol.
- The recoverable instances are significantly reduced in terms of time to recover all the failures, from 89.12 h to 25.28–23.59 h. With statistical significance, the total failures ($|F|$) and the total latency (z) are reduced by 312.33–319.30 failures and 2718.33–2761.79 h under the neuroplasticity-inspired scheduling protocols, and the preventability (P) of failures is increased by 0.077–0.080.
- The mean of total distance traveled by agents (Δ) is increased but not significantly, because agents keep traveling to handle disruptions, though the efficacy varies in all scenarios.
- The average latency (ζ) is about the same for all scheduling protocols. But as the ABP and ABP + AL significantly improve the preventability compared with FCFS, the total number of failures ($|F|$) is decreased and the total latency (z) is reduced significantly.

Table 6.12 Statistical comparisons of protocols in experiment with DLOC neuroplasticity-inspired scheduling protocols

Metric	Increase from FCFS to ABP[a]			Increase from FCFS to ABP + AL[a]		
	Average	SEM	p value**	Average	SEM	p value**
Total failures (|F|)	−312.33	43.71	0.000	−319.30	44.06	0.000
Total distances traveled by agents (Δ)	182.54	149.68	0.226	161.61	150.84	0.287
Total latency (z)	−2718.33	431.99	0.000	−2761.79	436.18	0.000
Preventability (P)	0.077	0.010	0.000	0.080	0.010	0.000

Metric	Increase from NN to ABP[a]			Increase from NN to ABP + AL[a]		
	Average	SEM	p value**	Average	SEM	p value**
Total failures (|F|)	−3.37	7.17	0.639	−10.34	8.69	0.237
Total distances traveled by agents (Δ)	−80.14	41.33	0.055	−101.07	49.87	0.045
Total latency (z)	−140.41	69.15	0.045	−183.87	81.23	0.026
Preventability (P)	−0.016	0.003	0.000	−0.013	0.003	0.000

[a]The increase from x to y is calculated as: subtract the metric of x from the metric of y

** The p-value is drawn from the one sample t-test on the hypothesis that the population mean equals 0. The hypothesis is rejected if the p value is less than the significance level (0.05)

- Compared with the NN policy, the developed DLOC neuroplasticity-inspired methods have significant advantage in minimizing the total latency (z). As shown in Table 6.12, the NN and the neuroplasticity protocols yield similar total number of failed elements ($|F|$). Therefore, the reduction in total latency (z) by neuroplasticity-inspired methods is contributed by the relatively less average latency for each failure (ζ). In Fig. 6.10b, starting from the 17th hours after the initial failures, the scenarios with ABP and ABP + AL protocols have less average failed elements at each time step. The preventability of NN policy, however, is higher than that of the neuroplasticity-inspired protocols.
- The observations indicate that, compared with neuroplasticity-inspired protocols, the NN policy spends relatively more time to prevent failures, and the delays reduce the chance to recover more failures within a short time.
- Compared with the ABP, the ABP + AL protocol is relatively more efficient in minimizing total failures ($|F|$) and total latency (z). This difference is indicated by the less increased average values and less p values when the metrics on the two protocols are compared with the metrics on the NN policy (see Table 6.12). Therefore, the AL protocol improves the response operations for the client networks that are able to inexpensively construct new links.

It can be concluded from the experiments that for the same client network, the CDR performance can be improved by the DLOC neuroplasticity-inspired scheduling protocols.

References

1. Bay Area Water Supply & Conservation Agency (BAWSCA). Hetch hetchy water system. http://bawsca.org/water-supply/hetch-hetchy-water-system/. Last retrieved July 2014
2. Bertsimas DJ, Van Ryzin G (1991) A stochastic and dynamic vehicle routing problem in the Euclidean plane. Oper Res 39(4):601–615
3. Chen XW, Nof SY (2012) Agent-based error prevention algorithms. Expert Syst Appl 39(1):280–287
4. Chen XW, Nof SY (2012) Conflict and error prevention and detection in complex networks. Automatica 48(5):770–778
5. Edan Y, Miles GE (1994) Systems engineering of agricultural robot design. IEEE Trans Syst Men Cybern 24(8):1259–1265
6. Hussein M, Fandi M, Muhtaseb J (2007) Effect of plant density on tomato yield and fruit quality growing in tuff culture. In: Proceedings of international symposium on fresh food quality standards: better food by quality and assurance, vol 741, pp 207–212
7. Lee S (2012) The role of centrality in ambulance dispatching. Decis Support Syst 54(1):282–291
8. Miller A, Allen PK (2004) Graspit!: a versatile simulator for robotic grasping. IEEE Robot Autom Mag 11(4):110–122
9. Pickett JA, Hamilton ML, Hooper AM et al (2010) Companion cropping to manage parasitic plants. Annu Rev Phytopathol 48:161–177
10. San Francisco Public Utilities Commission (SFPUC). Hetch Hetchy regional water system. http://sfwater.org/modules/showdocument.aspx?documentid=4192. Last retrieved July 2014
11. Shuang Q, Zhang M, Yuan Y (2014) Node vulnerability of water distribution networks under cascading failures. Reliab Eng Syst Safety 124:132–141

12. Surana A, Kumara S, Greaves M, Raghavan UN (2005) Supply-chain networks: a complex adaptive systems perspective. Int J Prod Res 43(20):4235–4265
13. Thames Water (2013) The process of repairing a leak. http://www.thameswater.co.uk/help-and-advice/11250.htm. Last retrieved Nov 2014
14. Watts DJ, Strogatz SH (1998) Collective dynamics of small-world networks. Nature 393(6684):440–442
15. Yagan O, Qian D, Zhang J, Cochran D (2012) Optimal allocation of interconnecting links in cyber-physical systems: interdependence, cascading failures and robustness. IEEE Trans Parallel Distrib Syst 23(9):1708–1721
16. Yeung BHB, Mills JK (2004) Design of a six DOF reconfigurable gripper for flexible fixtureless assembly. IEEE Trans Syst Men Cybern 34(2):226–235
17. Zhong H, Nof SY (2014) DLOC complex network model for supply network disruption response. In: Proceedings of international conference on production research—regional conference Europe, Africa and Middle East, Cluj-Napoca, Romania
18. Zhong H, Nof SY, Berman S (2015) Asynchronous cooperation requirement planning with reconfigurable end-effectors. Robot Comput Integr Manuf 34(8):95–104

Chapter 7
Evolving DLOC Theory and Emerging Applications

7.1 Machine Learning, Adaptive and Evolutionary DLOC

Machine learning can enable DLOC agents to improve and predict their future responses to disruptions based on their cumulative experience in their given environment; and lessons learned from previous collaborative interactions with other agents. Adaptive and bio-inspired evolutionary models and functions can further advance the agents' abilities for effective disruption response. Multi-agent systems with artificial intelligence learning, designed to manage disruptions for airline operations control [10] illustrate the value of learning for better disruption response.

The N2N service challenge is a race competition between the propagation of service requests in the client network and the limited service capability by the service team. The DLOC model developed in this research is a powerful analytical tool for improving the safe and secure continuity and resilience of e-Work. Emerging research of extending and applying the DLOC model will become a new and important subfield of Collaborative Control Theory (CCT).

The following three topics under the current research questions are worthwhile exploring.

1. **Team configuration**. It is necessary to add features that are specific to each individual client and server in the DLOC model. For instance, the direction of links in the client network can represent the flows of goods in a supply system, and nodes' weights in the server network can represent the capacity of production or delivery. Such additional features will change the objective function for team configuration and need further investigation.

   For the purpose of team reconfiguration as a disruption handling approach, best matching protocols [43] and resilience by teaming [6, 52] have been developed as useful techniques for disruption handling and control.
2. **Depot allocation**. Client network, such as future CPIs, will have active nodes that enable self-healing mechanisms [1]. The interaction and coordination of

© Springer Nature Switzerland AG 2020
H. Zhong and S. Y. Nof, *Dynamic Lines of Collaboration*,
Automation, Collaboration, & E-Services 6,
https://doi.org/10.1007/978-3-030-34463-4_7

external service agents with active nodes need further research. Moreover, the active clients may even assume the role of service agents during emergency to provide services to their passive peers [2]. Where to allocate service agents in collaboration with the active clients remains an open and interesting question [34, 35].

3. **Service schedule**. In the current research, all service operations are assumed to be error-free, which can be changed to a more realistic assumption. If errors and conflicts occur in N2N services, conflict resolution, re-scheduling of failed services, and error prevention require additional research to optimize the operations of servers. The re-planning of services is also related to the re-planning of the service team to remove erroneous servers. Application of the Join/Leave/Remain (JLR) principle of CCT, which addresses the analysis and decision to affiliate or dissociate from a networked system, is potentially useful to selectively update/reconfigure the service team [16, 64–66]. Extensive research on prognostics and prevention of errors and conflicts in e-Work networks [11–14] can also be useful for service-schedule revisions in future extensions of DLOC.

New research questions need to be answered as the DLOC and CCT research continues to explore new directions for solving emerging challenges. Examples of emerging research questions are as follows.

1. What are the criteria to design client network systems for better, more effective and responsive operations during an emergency disruption?
   The client network is assumed to be uncontrollable in the current research. However, the DLOC model is useful to examine whether the structure of the client network can help to reduce obstacles for service agents during emergency. For instance, future research can focus on optimizing the control structures of cyber-physical infrastructure with the objective of maintaining necessary connectivity while under disruptions.

2. How to use learning/evolving agents to improve the N2N services?
   The developed DLOC model imposes limitations as servers are modeled as having static capabilities in their provision of services. This assumption is acceptable for supporting one-time disruption response decisions. But it is more accurate and flexible to augment learning and evolving abilities in all or some of the service agents. Initial progress in this direction is reported on resilience informatics for disruption response in cyber-augmented networks, and collaborative response to disruption propagation in cyber physical systems and networks [45, 46]. Future research will also explore what to learn, how to learn, and the cost-effectiveness of learning.

7.2   Collaborative Visual Analytics and Decision Making with HUB-CI

As explained earlier, DLOC dynamically analyzes and recommends who the client agents that need support for disruption response actions are, and when and who the responding service agents are. Necessary tools that have emerged for these objectives are visual analytics and decision support.

Decision support systems (DSS) form a specific class of information systems which are meant to help the knowledge workers (managers placed at various management and control levels, experts, plant operators, and their respective agents) to solve complex problems that matter. The DSS input data may come from multiple sources: the decision-maker him/herself, various databases and networks, or, in the case of real-time applications, directly from cyber or physical sensors.

When managing large-scale systems, there are often more than one decision maker to consider all perspectives and aspects of decision problems. Group DSSs have been developed to enable such activities with distributed computer intelligence. The control mechanisms in groups are becoming more collaborative, rather than hierarchical, because collaborative control is more flexible, enabling robustness to major disturbances [19, 20]. A major trend in research is on how the Web is supporting more interactivity and collaboration with DSS. Organizations are building distributed, interactive decision support not only with virtual team structures, but also with their entire virtual organizations, based on the decision support technological platform [57].

Collaborative decision support is one instance of information systems that support e-Work by which all activities are modeled virtually on top of computers and networks [20, 47, 48]. They can contribute to sustainable advanced manufacturing [54, 55] and supply network decision making [67]. Based on CCT, several tools are being developed to improve cyber supported and cyber augmented systems collaboration [22, 24]. Collaborative intelligence (CI) is intelligence obtained from distributed sources, e.g., agents informing their local awareness, and the more global knowledge can boost an entity's ability for collaborating with other entities [17]. CI has been used in improving distributed and collaborative decision making with support of high-performance computing [71].

HUB-CI, a HUB for Collaborative Intelligence, is developed based on the HUBzero computing platform developed by Purdue University for scientific and professional collaboration in research and education [41]. This platform has supported several HUBs for collaborative work in different areas, e.g., nanoHUB.org for research and education about nanotechnology. Other cyber-supported collaboration systems are also available on the market. IBM produces JazzHub (hub.jazz.net) as a public hosted software development environment. HUBs with their advanced computing capabilities have centralized the effective and timely planning and control of distributed, shared resources, and have improved the collaboration of networked participants to share and to use knowledge and tools, particularly, newly developed knowledge.

For all HUBs and HUB-like systems, HUB-CI has been introduced to enable, improve and optimize the workflow and resource sharing collaboration by humans and automation systems [53, 70]. Examples of functions enabled by HUB-CI include:

1. *Collaborative network optimization* to select e-Service providers for knowledge-intensive tasks [17]. The support of computerized systems can reduce the human efforts in knowledge discovery and information matching for specific decision.
2. *Collaborative visualization* to improve shared understanding [72]. HUB-CI also matches participants in the system to proper tasks with proper interfaces. Then everyone involved in a decision process knows what to act and how to execute it in time-constrained and resource-constrained situations.

The features of HUB-CI allow participants to apply CI tools to make collaborative plans, designs, and decisions more efficiently [31]. The current research also applies CI to support the collaboration in N2N services.

An original co-insights framework for collaborative decision support has been designed and validated [69]. Better performance in decision making by a group of experts is attempted in modern industrial organizations with computational support. Collaborative intelligence is essential in this context to make collaborative decision support systems scalable. Collaboration is further facilitated by the sharing of insights by dynamic teams, with the online transfer of tacit knowledge. Making collaborative decision often consists of inter-related tasks, such as collecting information, analyzing data, generating solutions, synthesizing across the team, etc. It is a network-to-network challenge to determine how collaborative teams should be formed to fulfill the network of tasks.

A DLOC model has been introduced in this co-insight framework. The goal is to enable effective creation and sharing of collective and collaborative insights (co-insights) with automated supporting features. This research defines an alternative construct and solution for the team formation problem highlighting the smooth transfer of tacit knowledge in an organization. A Collaborative Agent Allocation Analysis (CA3) is designed for matching the task requirements and the expertise of participants. To validate CA3, experiments built on DLOC model are performed. CA3 significantly improves up to 35% of the matching scores, compared with conventional greedy task–participant matching methods. Thus, by better matching of tasks to participants, an effective and better sharing of insights is enabled for supporting collaborative decision making, accompanied with tacit knowledge transfer from experts to novices.

7.3 Heterogeneous, Collaborative Robots/Drones for Disruption Handling and Control

Emerging work on robots, drones, and integrated, collaborative cyber physical systems for monitoring, diagnostics, prevention, and mitigation of disruptions represent approaches that similar to DLOC must rely on dynamic lines of collaboration and command. Examples include:

- Teams of drones and autonomous cars for monitoring and rescue in smart cities [3, 28].
- Disaster rescue and disruption handling human-robot and robot-robot teams [33, 38–40, 44, 63].
- Collaborative mapping and monitoring sensor networks [26, 27, 30, 36, 37, 42]
- Human-robot collaborative systems for precision agriculture tasks that include MDR, Monitoring, Detection, and Response [4, 22, 56].

Emerging work relying on DLOC is described in [68]. In a fully automated industrial package loading and unloading scenario (which can be envisioned as handling emergency rescue supplies), a stream of uncertain types of packages needs to be continuously sorted and loaded onto designated destinations (trucks, shelves). For instance, a distribution center of an online store deploys robots to move goods from storages to delivery trucks [23]. To handle the variation of tasks, multiple types and combinations of robots, or heterogeneous robots including drones, are suitable for this task requirements. Packages may arrive in groups or batches due to their similar handling requirements (unit loads), or designated to similar destinations. Collaboration is required between the robots for handling large-scale items. It is a network-to-network problem: Design an efficient assignment protocol to dynamically allocate robot resources for the upcoming stream of package handling tasks.

Based on a DLOC model and intuitionistic fuzzy set theory, an adaptive collaborative task assignment protocol has been developed. The interactive collaboration between robots is dynamic to ensure the flexibility of the system. Experiments on simulated package handing systems, such as described here, indicate with statistical significance that the new approach shortens the total completion time, reduces total energy consumption, and increases loading accuracy, compared with the static collaborative task assignment. Further research is needed to include humans and robot teams in a cyber augmented work environment, where the designated package storage or loading equipment are also included by IoS and IoT [18].

7.4 Shared Quality Assurance in Manufacturing and Logistics

Production and supply networks are challenged in many dimensions, of which a prominent one concerns the quality of process, product, and logistics. A number of recent studies have addressed this problem in terms of disruption risk management [5], disruption recovery [25], and collaborative disruption handling [49].

A survey of data from over 900 companies, from over 60 countries [49] revealed four different patterns of disruption risk management commonly applied by manufacturing and logistics enterprises: (1) Passive; (2) Internal; (3) Collaborative; and

(4) Integral. The passive and internal approaches were found to be the least effective. The inter-organizational approach, combining collaborative and integral patterns were found to be the most effective. Moreover, they also influenced positively the internal continuity and security procedures. An overview of disruption risks, handling and control, including collaborative resolution approaches, and potential joint outcomes are summarized in Table 7.1.

An emerging approach for sharing quality assurance in a supply network based on DLOC is described by Candranegara et al. [9], as follows. In furniture manufacturing systems, machine/human errors and conflicts in and between operations can propagate and result in seriously inferior products. To eliminate conflicts/errors (CEs), quality assurance resources, e.g., inspectors and repair equipment, need to be allocated in the network of manufacturing workstations. As the repair resources require collaboration and services can be assumed to be interdepended, the challenge is a network-to-network problem. It can be solved by applying the DLOC model.

To allocate inspection resources to manufacturing workstations for efficient detection, prevention, and recovery of CEs, the workstations are ranked by the historical occurrence of CEs and the influence each given workstation have on other workstations, measured by eigenvector centrality [7, 8]. Two workflow scenarios are simulated: Inspection by humans, or by emerging autonomous systems. Experiments indicate that the developed method increases CEPD (Conflict and Errors Prevention and Detection) performance with statistical significance by reducing the time to completion, compared with the decentralized method (allocating resources to every workstation).

The DLOC based method also increases the preventability and reliability measures, while reducing the rectification and remedial cost, compared with the centralized method (allocating resources at the end of process). This case study validates the practical usefulness of the DLOC model in a manufacturing client network.

7.5 Security of Supply Networks

Supply networks include combination of digital supplies and physical supplies. Digital supplies are, for instance, signal flows through sensor networks, communications grids and networks, and computational services, such as client-server and cloud computing. Physical supplies flow through supply chains of goods, materials, and infrastructure services, such as power grids, water distribution, transportation, healthcare, and so on. With greater global supply interconnection, allied supply entities are increasingly more vulnerable to disruption risks [50–52].

As discussed in the previous section, it has been demonstrated that recovery from disruptions in those complex networks is slow, difficult and costly [5] (Revilla and Saenz 2015).

Resiliency of supply networks has emerged as a proactive way to design them, preparing for systematic approaches to train, prevent, and if necessary, mitigate,

Table 7.1 Disruption and its handling by global supply network entities (extended from [5])

Supply network interactions	Disruption risk characteristics	Collaborative disruption resolution	Joint outcomes
Norms: Flexibility Solidarity Information sharing Supply resource—sharing	Previous disruptions handling experience New disruption magnitude and risk New disruption potential propagation range	Collaborative disruption repair and recovery Collaborative problem solving Negotiating mutually satisfactory solutions Unresolved disruption; un- mitigated damage	Jointly creating supply value despite disruption Jointly creating supply value by innovations Jointly sharing value Alliance dissolution or restructuring
Progress: Planning Monitoring Incentives			
Trust: Process control Alliance protocols Quality standards Benevolence Competition[a]			

[a]Mixed strategy of cooperation and competition

handle and control disruptions that risk their security. Several particular approaches that can benefit from the DLOC model are as follows.

- Network structure analyses for disruption and resilience [29, 52], and resilient architecture design and planning based on them;
- Errors and conflicts prevention and response [11, 12, 14, 15, 32], and systematic risk assessment [21, 58] as a means to increase network resilience and security;
- Collaborative control theory applications for supply network security by task administration protocols [59–61], and sensor-based reconfiguration and control of cyber physical production systems [46, 62].

Can disruptions disappear from our world? No. But advanced methods to handle and control them can continue to evolve. The characteristics and advantages of DLOC and dynamic lines of collaboration, as discussed and illustrated in this book, can combine with the emerging research to understand better how to deal with disruptions: How to eliminate or minimize their risks; how to plan and train to mitigate them; and how to recover from them most effectively.

References

1. Amin M (2001) Toward self-healing energy infrastructure systems. Comput Appl Power 14(1):20–28
2. Ang CB, Nof SY (2001) Design issues for information assurance with agents: coordination in protocols and role combination in agents. CERIAS Technical Report, 2001-36, Purdue University
3. Basso M, Zacarias I, Tussi Leite C, Wang H, Pignaton de Freitas E (2018) A practical deployment of a communication infrastructure to support the employment of multiple surveillance drones systems. Drones 2(3):26
4. Bechar A, Wachs JP, Lumkes J, Nof SY (2012) Developing a human-robot collaborative system for precision agricultural tasks. In: 11th international conference on precision agriculture, Indianapolis, USA
5. Bello D, Bovell L (2012) Collaboration analysis: joint resolution of problems in global supply networks. Inform Knowl Syst Manage 11:77–99
6. Bhargava R, Reyes Levalle R, Nof SY (2016) A best-matching protocol for order fulfillment in re-configurable supply networks". Comput Ind 82:160–169
7. Bonacich P (1972) Factoring and weighting approaches to status scores and clique identification. J Math Sociol 2:113–120
8. Borgatti SP (2005) Centrality and network flow. Soc Networks 27(1):55–71
9. Candranegara G, Zhong H, Nof SY (2015) Conflict & error mgmt. based on collaborative control theory: a case study in the furniture industry. In: Proceedings of the 23rd international conference on production research
10. Castro AJM, Rocha AP (2017) Managing disruptions with a multi-agent system for airline operations control. In: Proceedings of advances in practical applications of cyber-physical multi-agent systems conference, Porto, Portugal, pp 307–310
11. Chen XW, Nof SY (2007) Prognostics and diagnostics of conflicts and errors over e-work networks. In: Proceedings of ICPR-19, Valparaiso, Chile, Aug 2007
12. Chen XW, Nof SY (2010) A decentralized conflict and error detection and prediction model. Int J Prod Res 48(16):4829–4843

13. Chen XW, Nof SY (2012) Agent-based error prevention algorithms. Expert Syst Appl 39(1):280–287
14. Chen XW, Nof SY (2012) Conflict and error prevention and detection in complex networks. Automatica 48(5):770–778
15. Chen XW, Landry SJ, Nof SY (2011) A framework of enroute air traffic conflict detection and resolution through complex network analysis. Comput Ind 62(8–9):787–794
16. Chituc CM, Nof SY (2007) The join/leave/remain (JLR) decision in collaborative networked organizations. Comput Ind Eng 53(1):173–195
17. Devadasan P, Zhong H, Nof SY (2013) Collaborative intelligence in knowledge based service planning. Expert Syst Appl 40(17):6778–6787
18. Dusadeerungsikul PO, Sreeram M, Nair A, Ramani K, Quinn A, Nof SY (2019) Collaborative requirement planning protocol for HUB-CI in factories of the future. In: Proceedings of ICPR-25, international conference on production research, Chicago
19. Filip FG (2008) Decision support and control for large-scale complex systems. Ann Rev Control 32:61–70
20. Filip FG, Zamfirescu CB, Ciurea C (2017) Computer-supported collaborative decision-making. Springer International Publishing, ACES Series (Automation, Collaboration, and E-Services), Berlin
21. Gjorgiev B, Antenucci A, Sansavini G, Volkanovski A (2018) A probabilistic risk assessment method for the security of supply in gas networks supported by physical models. In: Safety and reliability—safe societies in a changing world. CRC Press, Boca Raton, pp 1645–1653
22. Guo P, Dusadeeringsikul P, Nof SY (2018) Agricultural cyber physical system collaboration for greenhouse stress management. Comput Electron Agric 150:439–454
23. He Z, Aggarwal V, Nof SY (2018) Differentiated service policy in smart warehouse automation. Int J Prod Res 1–15
24. Hossain MS (2017) Cloud-supported cyber–physical localization framework for patients monitoring. IEEE Syst J 11(1):118–127
25. Ivanov D, Dolgui A, Sokolov B, Ivanova M (2017) Literature review on disruption recovery in the supply chain. Int J Prod Res 55(20):6158–6174
26. Jeong W, Nof SY (2009) A collaborative sensor network middleware for automated production systems. Comput Ind Eng 57(1):106–113
27. Jeong W, Ko H, Lim H, Nof S (2013) A protocol for processing interfered data in facility sensor networks. Int J Adv Manuf Technol 67(9–12):2377–2385
28. Johnsen SO (2018) Risks, safety and security in the ecosystem of smart cities. Risk Assessment, InTech
29. Kim Y, Chen YS, Linderman K (2015) Supply network disruption and resilience: a network structural perspective. J Oper Manage 33:43–59
30. Kim YD, Son GJ, Kim H, Song C, Lee JH (2018) Smart disaster response in vehicular tunnels: technologies for search and rescue applications. Sustainability 10(7):2509
31. Ko HS, Nof SY (2010) Design of collaborative e-service systems. In: Salvendy G, Karwowski W (eds.) Introduction to Service Engineering, pp. 227–252. Wiley, New York
32. Landry SJ, Chen XW, Nof SY (2013) A decision support methodology for dynamic taxiway and runway conflict prevention. Decis Support Syst 65(1):165–174
33. Lattanzi D, Miller G (2017) Review of robotic infrastructure inspection systems. J Infrastructure Syst 23(3):04017004
34. Lee S (2011) The role of preparedness in ambulance dispatching. J Oper Res Soc 62(10):1888–1897
35. Lee I, Sokolsky O, Chen S et al (2012) Challenges and research directions in medical cyber-physical systems. Proc. IEEE 100(1):75–90
36. Liu Y, Nof SY (2004) Distributed microflow sensor arrays and networks: design of architectures and communication protocols. Int J Prod Res 42(15):3101–3115
37. Liu Y, Nof SY (2008) Fault-tolerant sensor integration for micro flow-sensor arrays and networks. Comput Ind Eng 54(3):634–647

38. Liu Z, Suzuki T (2018) Using agent simulations to evaluate the effect of a regional BCP on disaster response. J Disaster Res 13(2):387–395
39. Liu Y, Gao J, Zhao J, Shi X (2018) A new disaster information sensing mode: using multi-robot system with air dispersal mode. Sensors 18(10):3589
40. Matsuno F, Sato N, Kon K, Igarashi H, Kimura T, Murphy R (2014) Utilization of robot systems in disaster sites of the great eastern Japan earthquake. In: Field and service robotics, pp 1–17
41. McLennan M, Kennell R (2010) HUBzero: a platform for dissemination and collaboration in computational science and engineering. Comput Sci Eng 12(2):48–52
42. Michael N, Shen S, Mohta K, Kumar V, Nagatani K, Okada Y, Kiribayashi S, Otake K, Yoshida K, Ohno K, Takeuchi E (2014). Collaborative mapping of an earthquake damaged building via ground and aerial robots. In: Field and service robotics, pp 33–47
43. Moghaddam M, Nof SY (2017) Best matching theory & applications. Springer ACES Series (Automation, Collaboration, and E-Services), Berlin
44. Murphy RR (2014). Disaster robotics. MIT Press, Cambridge
45. Nguyen WPV, Nof SY (2018) Resilience informatics for cyber-augmented manufacturing networks (CMN): centrality, flow, and disruption. Stud Inform Control 27(4):377–384
46. Nguyen WPV, Nof SY (2019) Collaborative response to disruption propagation (CRDP) in cyber-physical systems and complex networks. Decis Support Syst 117:2019
47. Nof SY (2003) Design of effective e-work: review of models, tools, and emerging challenges. Prod Plan Control 14(8):681–703
48. Nof SY, Silva JR (2018) Perspectives on manufacturing automation under the digital and cyber convergence (invited). Polytechnica 1:36–47
49. Revilla E, Saenz MJ (2017) The impact of risk management on the frequency of supply chain disruptions: a configurational approach. Int J Oper Prod Manage 37(5):557–576
50. Reyes Levalle R (2018) Resilience by teaming in supply chains and networks. Springer ACES Series (Automation, Collaboration, and E-Services), Berlin
51. Reyes Levalle R, Nof SY (2015) Resilience by teaming in supply network formation and re-configuration. Int J Prod Econ 160:80–93
52. Reyes Levalle R, Nof SY (2017) Resilience in supply networks: definition, dimensions, and levels. Ann Rev Control 43:224–236
53. Seok H, Nof S (2011) The HUB-CI initiative for cultural, education and training, and healthcare networks. In: 21st International Conference on Production Research, Stuttgart, Germany
54. Seok H, Nof SY, Filip FG (2012) Sustainability decision support system based on collaborative control theory. Ann Rev Control 36(1):85–100
55. Seok H, Kim K, Nof SY (2016) Intelligent contingent multi-sourcing model for resilient supply networks. Expert Syst Appl 51:107–119
56. Serrano D, Astolfi P, Bardaro G, Gabrielli A, Bascetta L, Matteucci M (2017) GRAPE: ground robot for vineyArd monitoring and protEction. ROBOT 2017: third Iberian Robotics Conference, vol 1, p 249
57. Shim J, Warkentin M, Courtney JF et al (2002) Past, present, and future of decision support technology. Decis Support Syst 33(2):111–126
58. Su H, Zhang J, Zio E, Yang N, Li X, Zhang Z (2018) An integrated systemic method for supply reliability assessment of natural gas pipeline networks. Appl Energy 209:489–501
59. Tkach I, Edan Y, Nof SY (2011) A framework for automatic multi-agents collaboration in target recognition tasks. In: Proceedings of 21st ICPR, Stuttgart, Germany, Aug 2011
60. Tkach I, Edan Y, Nof SY (2012) Security of supply chains by automatic multi-agents collaboration. In: Proceedings of INCOM, 14th IFAC symposium on information control problems in manufacturing, Bucharest, Romania, May 2012
61. Tkach I, Edan Y, Nof SY (2017) Multi-sensor task allocation framework for supply networks security using task administration protocols. Int J Prod Res 55(18):5202–5224
62. Tomiyama T, Moyen F (2018) Resilient architecture for cyber-physical production systems. CIRP Ann Manuf Technol 67(1):161–164
63. Whitman J, Zevallos N, Travers M, Choset H (2018) Snake robot urban search after the 2017 Mexico City earthquake. In: 2018 IEEE international symposium on safety, security, and rescue robotics (SSRR), pp 1–6

64. Yoon SW, Nof SY (2010) Demand and capacity sharing decisions and protocols in a collaborative network of enterprises. Decis Support Syst 49(4):442–450
65. Yoon SW, Nof SY (2011) Affiliation/dissociation decision models in demand and capacity sharing collaborative network. Int J Prod Econ 130(2):135–143
66. Yoon SW, Nof SY (2011) Cooperative production switchover coordination for the real-time order acceptance decision. Int J Prod Res 49(6):1813–1826
67. Yoon SW, Velasquez JD, Ko HS, Chen X, Nof SY (2010) Collaborative distributed-training control system for transportation and emergency response. In: Proceedings of IIE Annual Conference and Expo, Cancun, Mexico, May 2010
68. Zhang L, Zhong H, Nof SY (2015) Adaptive fuzzy collaborative task assignment for heterogeneous multi-robot systems. Int J Intell Syst 30(6):731–762
69. Zhong H (2016). Dynamic lines of collaboration in e-work systems. Unpublished doctoral dissertation, School of Industrial Engineering, Purdue University, West Lafayette
70. Zhong H, Nof SY (2013) Collaborative design for assembly: the HUB-CI model. In: Proceedings of the 22st international conference on production research, Iguassu Falls, Brazil
71. Zhong H, Reyes Levalle R, Moghaddam M, Nof SY (2015) Collaborative intelligence—definition and measured impacts on internetworked e-Work. Manage Prod Eng Rev 6(1):67–78
72. Zhong, H., Nof, S.Y., Ozsoy, E. (2015b) Co-Viz: matching tools in collaborative visual analytics. In: Proceedings of industrial and systems engineering research conference, Nashville, USA

Glossary

ACRP Asynchronous Cooperation Requirement Planning
AL Auxiliary Links
ABP Activity-Based Priority
BAWSCA Bay Area Water Supply & Conservation Agency
BA Barabási-Albert scale-free network
CA3 Collaborative Agent Allocation Analysis
CBA Centrality-Based depot Allocation
CCT Collaborative Control Theory
CDR Collaborative Disruption Response
CDS Collaborative Decision Support
CEPD Conflict and Errors Prevention and Detection
CER Conflict Error Resolution
CI Collaborative Intelligence
CLM Collaboration Lifecycle Management
CPI Cyber-Physical Infrastructure
CPS Cyber-Physical System
CRP Collaboration Requirement Planning
DDoS Distributed Denial of Service
DHS U.S. Department of Homeland Security
DLOC Dynamic Lines of Collaboration
EIA U.S. Energy Information Administration
EMC E-Metrics and e-Criteria
EMS Emergency Medical Service
ER Erdős-Rényi random graph
FTT Fault-Tolerance by Teaming
GQO Grasp Quality Optimization
HH Hetch Hetchy water system
HUB-CI HUB-based Collaborative Intelligence
JLR Join/Leave/Remain decisions
LOC Line of Collaboration and Communication

© Springer Nature Switzerland AG 2020
H. Zhong and S. Y. Nof, *Dynamic Lines of Collaboration*,
Automation, Collaboration, & E-Services 6,
https://doi.org/10.1007/978-3-030-34463-4

MDR Monitoring, Detection, and Response/Recovery

N2N Network-to-Network

NN Nearest Neighbor

NP Nondeterministic Polynomial time

OOR Order of Reconfigurability

QoS Quality of Service

RDA Random Depot Allocation

REE Reconfigurable End-Effectors

SFPUC San Francisco Public Utilities Commission

SPT Shortest Processing Time

TIE Teamwork Integration Evaluator

TRP Traveling Repair-agent Problem

WS Watts-Strogatz small-world model

WDN Water Distribution Network

A Service agent

$btw(\cdot)$ The function to calculate betweenness centrality

C Collaboration-ability in R

CASCAD Cascading function of services

C_{config} Estimated total reconfiguration cost

$c_{config}(i)$ Reconfiguration cost in the ith layer of the configuration network

C_{grasp} Minimum total grasp cost

$c_{grasp}(j, \delta)$ Grasp cost of object with dimension δ using configuration j

c_{ij} The weight of the links from i to j, $i, j \in D \cup F$

D The set of depots

$d(i, j)$ Distance function between $i, j \in D \cup F$

$deg_R(\cdot)$ Degree function in R

E Inter-edge between G and R

e_i Number of links between the neighbors of node i

F The set of disruptions

$\tilde{F}$ The set of disruptions without response mechanisms

$|F_{max}|$ Max cascade

F_t The set of disrupted elements at time t

f The frequency of switching collaboration configurations

G A client network

i, j, k, a, b Element of a set

(i,j) A link connecting node i and node j

J Total design cost of REE

L Links in G

$l_{neighbor}(j, i)$ Number of links between node j and node i's neighbors

M Total number of configurations

M_i Number of reconfigurable parts in layer i

$\mathbb{M}_i$ Maximum number of reconfigurable parts in layer i

N Nodes in G

n_A Number of nodes in A

N_{arm} Number of arms in an automated harvesting system

$\mathbb{N}_{\mathbf{arm}}$ Maximum number of arms

$N_{(k)}(i)$ Neighbors of the kth order for node i

$N_{\mathbf{T}}$ Target nodes for auxiliary links

$n_{\mathbf{R}}$ Number of replications

P_G Degree distribution of G

$Pr(\cdot)$ Probability of an event

$P_{\mathbf{recover}}$ Recoverability

$P_{\mathbf{prevent}}(t)$ Preventability at time t

q Average grasp quality

R A network representation of a service team

S Any subset of F

t Time variable

$t_{\mathbf{D}}$ Travel time or reconfiguration time of agent

$t_{\mathbf{max}}$ Simulation length

$t_{\mathbf{repair}}$ Repair timespan

$t_{\mathbf{recover}}$ Recovery time

u Service propagation speed

$U(\cdot)$ Uniform distribution

v Agent's traveling velocity

x_{ij} Decision variable of whether an agent goes from i to j, $i, j \in D \cup F$

Y Total yield in automated harvesting

z DLOC objective function, total latency

$\beta_r, \beta_\rho, \beta_g$ Normalization factors for total design cost J

γ_i The rupture reference of disruption i

δ Dimension of the target object

Δ Total travel distance by service agents

ε Homeostatic point in neuron activity

ζ Average latency

η Minimum activation threshold in neuron activity

θ Connectivity

λ Fruit density

μ Service rate of harvesting process

υ_i The location of the disruption i

ρ Estimated production cost

σ_i Time timestamp when disruption i is repaired

$\sigma_{\mathbf{cluster}}(i)$ Cluster coefficient of node i

τ_i Initial timestamp of disruption i

φ Cascading threshold

Ψ Simulation scenario

ω Service priority

Uncited References

1. Adam N (2010) Workshop on future directions in cyber-physical systems security. U.S. Department of Homeland Security
2. Bellocci T, Lehto MR, Nof SY (2003) Assuring information quality in industrial enterprises: experiments in an ERP environment. In: Proceedings of the 10th International Conference on Human-Computer Interaction, Crete, Greece, pp. 654–658
3. Benini L, Farella E, Guiducci C (2006) Wireless sensor networks: enabling technology for ambient intelligence. Microelectron J 37(12):1639–1649
4. Brailer DJ (2005) Interoperability: the key to the future health care system. Health Affairs 24:W5-19–W5-211
5. Burmester M, Magkos E, Chrissikopoulos V (2012) Modeling security in cyber-physical systems. Int J Crit Infrastructure Protect 5:118–126
6. Carver L, Turoff M (2007) Human-computer interaction: the human and computer as a team in emergency management information systems. Commun ACM 50(3):33–38
7. Chartuverdi AR, Hutkinson GK, Nazareth DL (1993) Supporting complex real-time decision-making through machine learning. Decis Support Syst 10:213–233
8. Decker S, Schoenberg T (2011) Rolls-Royce loses engine-patent case against United Technologies' Pratt. Bloomberg, http://www.bloomberg.com/news/2011-05-20/rolls-royce-loses-jet-engine-case-against-pratt-whitney-1-.html. Last retrieved 25 Sep 2013
9. Demetrius L, Manke T (2004) Robustness and network evolution—an entropic principle. Physica A 346(3–4):682–696
10. Delic K, Douillet L, Dayal, U (2001) Towards an architecture for real-time decision support systems: challenges and solutions. In: International symposium on database engineering and applications. Grenoble, France, pp 303–311
11. DHS, U.S. (2011) national preparedness goal. http://www.fema.gov/media-library-data/20130726-1828-25045-9470/national_preparedness_goal_2011.pdf
12. EIA, U.S. (2014) Table B.1 major disturbances and unusual occurrences, year-to-date. Electric Power Monthly. www.eia.gov/electricity/monthly/epm_table_grapher.cfm?t=epmt_b_1. Last retrieved: Oct 2014)
13. Filip FG (1995) Towards more humanized real-time decision support systems. In: Camarinh-Matos LM et al (eds) Balanced automation systems. Springer Science + Business Media, Dordrecht
14. Filip FG, Donciulescu DA, Filip CI (2002). Towards intelligent real-time decision support systems for industrial milieu. Stud Inform Control 11(2):303–311
15. Filip FG, Leiviska K (2009) Large-scale complex systems. In: Nof SY (ed) Springer handbook of automation. Springer, Dordrecht, pp 619–638

© Springer Nature Switzerland AG 2020
H. Zhong and S. Y. Nof, *Dynamic Lines of Collaboration*,
Automation, Collaboration, & E-Services 6,
https://doi.org/10.1007/978-3-030-34463-4

16. Jackson T, Austin J, Fletcher M et al (2005) Distributed health monitoring for aero-engines on the grid. In: DAME, IEEE aerospace conference, pp. 3738–3747

17. Juels A, Oprea A (2013) New approaches to security and availability for cloud data. Commun ACM. 56(2):64–73

18. Kemmerer RA, Vigna G (2002) Intrusion detection: a brief history and overview. Computers 35(4):27–30

19. Ko HS, Nof SY (2012) Design and application of task administration protocols for collaborative production and service systems. Int J Prod Econ 135:177–189

20. Ko H, Lim H, Jeong W, Nof SY (2010) A statistical analysis of interference and effective deployment strategies for facility specific wireless sensor networks. Comput Ind 61:472–479

21. Ko HS, Yoon SW, Nof SY (2011) Intelligent alert systems for error and conflict detection in supply networks. In: Proceedings of IFAC Congress, Milan, Italy, Aug 2011

22. Loomis E, Walrath J (2008) Collaborative decision support for layered sensor webs. In: IEEE National Aerospace and Electronics Conference, Dayton, OH, pp. 1–5

23. Lorincz K, Malan D, Fulford-Jones T, Nawoj A, Clavel A, Shnayder V et al (2004) Sensor networks for emergency response: challenges and opportunities. IEEE Pervasive Comput 16–23

24. Naedele M (2007) Addressing IT security for critical control systems. In: 40th Hawaii international conference on system science (HICSS-40)

25. Nair A, Vidal JM (2011) Supply network topology and robustness against disruptions—an investigation using multi-agent model. Int J Prod Res 49(5):1391–1404

26. Nof SY (1994) Information and collaboration models of integrations. Kluwer Academic Publishers, Dordrecht

27. Nof SY (2017) Collaborative control theory and decision support systems. Comput. Sci. J. Moldova 25(2)

28. Nof SY, Cheng GJ, Weiner AM, Chen XW, Bechar A, Jones MG, Reed CB, Donmez A, Weldon TD, Bermel P, Bukkapatnam SST, Cheng C, Kumara SRT, Bement A, Koubek R, Bidanda B, Shin YC, Capponi A, Lee S, Lehto MR, Liu AL, Nohadani O, Dantus M, Lorraine PW, Nolte DD, Proctor RW, Sardesai HP, Shi L, Wachs JP, Zhang XC (2013) Laser and photonic systems integration: Emerging innovations and framework for research and education. Human Factors Ergon Manuf Serv Ind 23(6):483–516

29. Pathak SD, Dilts DM, Biswas G (2007) On the evolutionary dynamics of supply network topologies. IEEE Trans Eng Manage 54(4):662–672

30. Pereira CE, Carro L (2007) Distributed real-time embedded systems: recent advances, future trends and their impact on manufacturing plant control. Ann Rev Control 31(1):81–92

31. PwC (2013) 2013 patent litigation study: big cases make headlines, while patent cases proliferate. Last retrieved 16 Oct 2013. http://www.pwc.com/en_US/us/forensic-services/publications/assets/2013-patent-litigation-study.pdf

32. Sampigethaya K, Poovendran R (2013) Aviation cyber-physical system: foundations for future aircraft and air transport. Proc IEEE 101(8):1834–1855

33. Shafiq B, Joshi JBD, Bertino E, Ghafoor A (2005) Secure interoperation in a multidomain environment employing RBAC policies. IEEE Trans Knowl Data Eng 17(11):1557–1577

34. Sheyner O, Wing J (2004) Tools for generating and analyzing attack graphs. In: de Boer et al FS (ed) Formal methods for components and objects. Springer, Berlin, pp 344–371

35. Skorobogatov S, Woods C (2012) Breakthrough silicon scanning discovers backdoor in military chip. In: Proceedings of 14th international workshop cryptographic hardware and embedded systems, Leuven, Belgium, pp 23–40

36. Sridhar S, Hahn A, Govindarasu M (2012) Cyber-physical system security for the electric power grid. Proc IEEE 100(1):210–224

37. Symantec (2013) Vulnerability trends—SCADA vulnerabilities. Last retrieved 17 Oct 2013. http://www.symantec.com/threatreport/topic.jsp?id=vulnerability_trends&aid=scada_vulnerabilities

38. Sztipanovits J, Koutsoukos X, Karsai G (2012) Toward a science of cyber-physical system integration. Proc IEEE 100(1):29–44

39. Thadakamalla HP, Raghavan UN, Kumara S, Albert R (2004) Survivability of multiagent-based supply networks: a topological perspective. IEEE Intell Syst 19(5):24–31
40. Turoff M (2002) Past and future emergency response information systems. Commun ACM 45(4):29–32
41. Verl A, Lechler A, Schlechtendahl J (2012) Glocalized cyber physical production systems. Prod Eng 6(6):643–649
42. Wu F, Kao Y, Tseng Y (2011) From wireless sensor network towards cyber-physical systems. Pervasive Mob Comput 7(4):397–413
43. Zhong H (2012) HUB-based robotics. MS thesis, School of Industrial Engineering, Purdue University
44. Zhong H, Wachs JP, Nof SY (2013) A collaborative telerobotics network framework with hand gesture interface and conflict prevention. Int J Prod Res 51(15):4443–4463

Printed in the United States
By Bookmasters